Food Engineering Series

T0185137

Series Editor
Gustavo V. Barbosa-Cánovas, Washington State University

Advisory Board
José Miguel Aguilera, Pontificia Universidad Católica de Chile
Xiao Dong Chen, Monash University
J. Peter Clark, Clark Consulting
Richard W. Hartel, University of Wisconsin
Albert Ibarz, University of Lleida
Jozef Kokini, Rutgers University
Michèle Marcotte, Agriculture & Agri-Food Canada
Michael McCarthy, University of California at Davis
Keshavan Niranjan, University of Reading
Micha Peleg, University of Massachusetts
Shafiur Rahman, Sultan Qaboos University
M. Anandha Rao, Cornell University
Yrjö Roos, University College Cork
Walter L. Spiess, Bundesforschungsanstalt
Jorge Welti-Chanes, Universidad de las Américas-Puebla

For further volumes:
http://www.springer.com/series/5996

J. Peter Clark

Case Studies in Food Engineering

Learning from Experience

J. Peter Clark
644 Linden Avenue
Oak Park IL 60302
USA
jpc3@worldnet.att.net

ISBN 978-1-4899-8424-1   ISBN 978-1-4419-0420-1 (eBook)
DOI 10.1007/978-1-4419-0420-1
Springer Dordrecht Heidelberg London New York

© Springer Science+Business Media, LLC 2009
Softcover re-print of the Hardcover 1st edition 2009
All rights reserved. This work may not be translated or copied in whole or in part without the written
permission of the publisher (Springer Science+Business Media, LLC, 233 Spring Street, New York,
NY 10013, USA), except for brief excerpts in connection with reviews or scholarly analysis. Use in
connection with any form of information storage and retrieval, electronic adaptation, computer software,
or by similar or dissimilar methodology now known or hereafter developed is forbidden.
The use in this publication of trade names, trademarks, service marks, and similar terms, even if they are
not identified as such, is not to be taken as an expression of opinion as to whether or not they are subject
to proprietary rights.

Printed on acid-free paper

Springer is part of Springer Science+Business Media (www.springer.com)

Preface

Purpose and Background of the Book

I write this book as a life-long student and practitioner of food engineering who realizes that relatively little of what I have learned to do in process development and food plant design was taught to me, or anyone, in a conventional academic course. Rather, I and others who practice in industry or are consulting have learned by experience and, frankly, by making mistakes. Once I gave a plenary lecture at an International Conference on Engineering and Food (ICEF 7) (Clark 1997b) and a friend commented that all my examples sounded like failures. Well, that was not quite true, but I also feel that we do primarily learn from our mistakes – unfortunately it is rare that great successes are dissected.

This book has several purposes:

1. To serve as a source of information about a representative collection of food processes with which I have had experience;
2. To convey some practical lessons about process development and plant design; and
3. To serve educators as a resource for class problems and discussion.

The book is organized in three broad parts. The first concerns processes that are primarily physical, such as mixing. The second concerns processes that also involve biochemical changes, such as thermal sterilization. The third part addresses some broader issues that I have not seen discussed elsewhere, including how to tour a plant, how to choose among building a new plant, expanding or renovating, and how to develop processes.

Entire volumes have been written by others on many of the processes discussed here, so I am not attempting to supply the definitive description of each. Rather I intend to provide my own understanding in sufficient detail to set the context in which the various lessons can be illustrated.

I should also mention that I have tried very hard to avoid disclosing proprietary information and I have not intentionally revealed the names of specific clients or sources of information. My career spans over 40 years and so some experiences are relatively well aged while others are much more recent. Some companies consider the fact that they used my services (or those of the firms for whom I worked) as confidential; others do not, but I have written as if all felt that way. Some of my past clients probably consider some of our mutual developments as proprietary, and over the years I was trusted with some very sensitive information, such as product formulas. I believe that the facts I am using are in the public domain.

I have limited illustrations to those I felt are essential. It is a useful exercise for the reader to locate on the Internet and in other sources additional illustrations if interested.

Some of the material in this book previously appeared in a different form in columns on processing since 2002 in *Food Technology* magazine, published by the Institute of Food Technologists. I am grateful to Neil Mermelstein, Bob Swientek, Roy Hlavacek, and Mary Ellen Kuhn, editors and publisher who gave me the opportunity to write those columns and helped to make them better.

I dedicate this book with love and gratitude to my wife since 1968, Nancy, who read each draft, made cogent suggestions, and has supported this effort and my entire career generously and wisely.

Oak Park, Illinois   J. Peter Clark

Contents

About the Author

Peter Clark is a consulting process engineer in Oak Park, IL, and Contributing Editor of Food Technology magazine, with a monthly column on processing since 2002. Previously, he was Vice President, Technology with Fluor Daniel (a major engineering firm); Senior Vice President, Process Technology of A. Epstein and Sons International (a large architecture and engineering firm with a food practice), and President of Epstein Process Engineering (a subsidiary); Director of Research for ITT Continental Baking Company; Assistant, then Associate Professor of Chemical Engineering at Virginia Tech; and he began his career as a Research Engineer with the US Agricultural Research Service.

He earned his B.S.Ch.E from Notre Dame and his Ph.D. from the University of California, Berkeley. He has published over 40 papers and seven books in addition to his columns. He is a registered professional engineer in Illinois and Virginia, is a Fellow of AIChE, and a past chair of IFT Food Engineering Division and the New Products and Technologies Committee and of AIChE Food, Pharmaceutical and Bioengineering Division, from which he received the Division Award in 1998.

Chapter 1
Introduction

Case studies typically are self-contained vignettes that are intended to teach one or more lessons, which are meant to be discovered by students or readers through discussion and reflection. Many teachers have found them to be useful pedagogical devices. In my personal experience, cases and examples provide the motivation to investigate underlying theory and principles. Using a specific example not only is intrinsically interesting but also introduces painlessly the need to understand various aspects of science and engineering. Collections of case studies have been assembled for education in business management, law, and ethics, among others, perhaps most famously at Harvard Business School. To my knowledge, there has not previously been published such a collection for food engineering.

Food engineering has become a reasonably well-recognized specialization, but there are still relatively few formal academic programs that award a degree in the field. Those who practice food engineering are often educated in chemical, mechanical, or agricultural and biosystems engineering. They may or may not have studied food science. Many food science and technology academic programs require at least one course in engineering, during which educators attempt to convey as many of the basic principles of engineering as they can. This book is intended to be used in such courses or in food engineering electives offered by other departments. It should also be useful to engineers and scientists new to the food industry and to others who want to acquire a general understanding of the technical aspects of the industry.

1.1 How to Use Case Studies

This is not so much a text as much as it is a resource. At various points I will try to suggest potential problems or exercises, usually derived from calculations I may have made in real life. An instructor can create numerous variations on many of the cases I describe simply by changing some values, such as capacity or physical properties, if he or she wishes.

It might be instructive to use some cases as a foundation for small group discussions, considering such issues as alternative solutions, debating the benefits and costs of chosen paths, and deriving additional lessons from the facts given.

J.P. Clark, *Case Studies in Food Engineering*, Food Engineering Series, DOI 10.1007/978-1-4419-0420-1_1, © Springer Science+Business Media, LLC 2009

After using cases in my own academic time and in teaching short courses, I have concluded that one of the best things an instructor or leader can do is keep his mouth shut. Let the participants articulate their own opinions, prodding them as necessary to stimulate discussion. Some of the best prods start with "Why?" or "How?". Tolerate silence – the students will fill it. A leader can share his or her own opinions, but usually after the team has had its chance.

I imagine that this book might supplement conventional texts in food engineering or food science in which the fundamental tools are taught, but which often lack real-world application.

Above all, I hope that by conveying some of the lessons I have learned, I will spare the reader a few of the bumps and bruises I have experienced.

There is a famous story of a ship captain interviewing candidates to be his river pilot in an unfamiliar part of the world. The first two candidates bragged about how they had never struck a rock or shoal. The third admitted that he had probably hit every hazard on the river. The third pilot got the job because the captain understood that he really knew the river while the others had just been lucky. May this book be something like a map, drawn by that third pilot, for those just starting to navigate the challenges of the food industry.

Part I
Processes Based Largely on Physical Operations

Chapter 2
Dry Mixing

Mixing of powders is a very common unit operation in many industries, including food processing. Oddly, it is rarely discussed in undergraduate engineering classes and is not a common subject of academic research. Nonetheless, it is at the heart of many food plants either to produce an end product, such as dry soup mix, or to make an intermediate material, such as a topical seasoning for snacks (Clark 2005a, 2007).

There are many choices for mixing equipment, including horizontal and vertical agitated chambers, tumbling vessels, and air agitated operations. Mixing can be continuous or batch, but delivering the components and removing the product are often more critical than the choice of mixing equipment.

Some of the key issues in solids mixing are as follows:

- material handling,
- proper mix time,
- mixer volume,
- scheduling and surge,
- segregation, and
- feeding, especially in the case of continuous mixing.

Material handling. This refers to the delivery, scaling, and conveying of various components of a mixture, typically solids such as flour, sugar, salt, or other ingredients. Most formulas have major, minor, and micro ingredients, distinguished by their relative amounts in the formula. The dividing points are matters of judgment, but one approach is to say that amounts >10% are major, 1–10% minor, and <1% micro ingredients. Micro ingredients, while occurring in relatively small amounts, are often critical to the functionality of the final mix, as they may be vital nutrients, leavening agents, colors, or flavors. Thus, their proper dispersion in the mix may be especially crucial.

Each category of ingredient may be delivered in a different way. For example, major ingredients are often delivered in bulk, unloaded pneumatically into bins, and delivered to a mixer by pressure or vacuum pneumatic lines. Pneumatic transfer refers to conveying solids in pipes or tubes by suspending the solid particles in moving air. Depending on the concentration of solids in air, the transfer may be considered dense phase or dilute phase.

J.P. Clark, *Case Studies in Food Engineering*, Food Engineering Series,
DOI 10.1007/978-1-4419-0420-1_2, © Springer Science+Business Media, LLC 2009

Mix time. Mixing is a case where more is not necessarily better. There is usually an optimal mix time, which must be determined experimentally. The experiment is tedious because mixing is determined by measuring the standard deviation of some critical component. This requires taking multiple samples, at least ten, from various parts of the mixer at a succession of times. Often, mixing times are determined by using an easy-to-analyze component, such as salt, but care must be taken that the results apply to the material of most interest, since it may have different particle size and density than salt does. If mix time is determined on small-scale equipment, scale-up parameters can be established by using similar geometric ratios and keeping the Froude number constant. The Froude number is a dimensionless group equal to N^2D/g, where N is rotational speed, D is a characteristic dimension of the mixer such as diameter, and g is the acceleration due to gravity, all in consistent units. The implication of this approach is that as a mixer of the same geometric ratio, such as length to diameter, gets larger (i.e., the diameter gets larger), the required rotational speed is reduced to keep the Froude number constant. The resulting mix time in a larger mixer might actually increase because the intention is to keep the number of turnovers Nt, where t is the mix time, constant (Valentas et al. 1991).

Rotation speed also matters and usually has an optimum value. At too low a speed, there is inadequate agitation; but at low speeds, avalanching flow can occur, which is efficient in mixing. At too high a speed, centrifugal force sends all the particles to the perimeter. The Froude number is the relevant scaling parameter:

$$F_r = N^2D/g \qquad (2.1)$$

where

> N is rotation speed, s^{-1}
> D is radius of cylinder (or another characteristic dimension of the equipment), m
> g is acceleration of gravity, m/s^2
> $F_r = 1$ defines the beginning of centrifuging, so values considerably less than 1 are needed for mixing (e.g., 0.03).

Once a process study has been performed, it is important to control mix times because excessive mixing can promote segregation. Furthermore, unless all subsequent steps are carefully designed, a uniform mix can become segregated just by falling through chutes or filling bins.

Mixer volume. Most solids mixers have working volumes equal to 50 % of their total volume. This means, for instance, that a ribbon or paddle blender should only be filled to just above its shaft. Other designs, such as tumblers and V-blenders, likewise must be only partially filled. This can be a source of friction between operators who want to maximize batch sizes and developers who understand the limitations of the equipment. If a mixer is overfilled, it is difficult for the solids to be moved by the agitator and mixing will be poor at best. Net production is actually improved by using equipment properly because mix times will be shorter and quality higher, resulting in less rework.

Fig. 2.1 Simple mixer
(symbol for mixer with feed
and discharge)

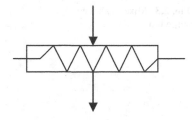

Scheduling and surge. The simplest mixing process consists of loading directly into a mixer, mixing for the correct time, and then dumping directly into packages or processing further. This requires some elevation of the mixer to allow room for packaging or conveying under the equipment. Often, there is a work platform on which bagged ingredients are placed and an operator manually loads the mixer. The mixer is idle during loading and unloading (Fig. 2.1).

A modest improvement in cycle time is achieved by delivering some ingredients in bulk to a receiver and scale, but this requires more head room and additional equipment (Fig. 2.2).

The next level of complexity adds a receiving bin, into which the components of a formula are loaded, manually through a bag dump station or pneumatically, while the previous batch is being mixed. The mixer is idle during unloading but is loaded quickly from the receiving bin once it is empty. The assumption is that successive batches of the same formula or of compatible formulas are being made. If cleaning is required between batches, then that time is added to the cycle (Fig. 2.3).

An obvious further improvement is to add a holding bin after the mixer. Now the mixer has minimal idle time, so production throughput, all other things being equal, is maximized. However, there is a price in both space and capital cost. Because solids transfers rely on gravity, this configuration is vertical and may not fit in an existing building (Fig. 2.4).

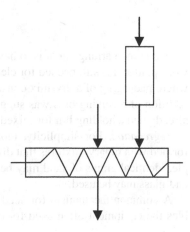

Fig. 2.2 Mixer with feed bin
and direct feed

Fig. 2.3 Mixer with feed surge bin

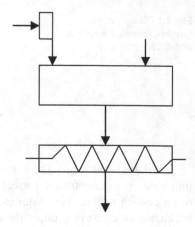

Fig. 2.4 Mixer with feed and discharge surge bins

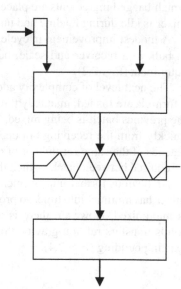

A common arrangement is to have bag dumps and controls at ground level, but a work platform is still needed for cleaning and maintenance. In at least one instance, where packaging of a dry-mix consumer product was proposed to be increased, the addition of receiving bins was suggested to enable an existing mixer to keep up. It already had a holding bin for mixed product.

Segregation. For simplicity, most research on segregation and mixing is performed with binary mixtures that differ in size or density, but rarely both. For example, identical grains of sand may be dyed two colors. Or same-size spheres of steel and glass may be used.

A common mechanism for academic study is a rotating cylinder, which resembles the equipment often used for coating particles or making cement. Sometimes the cylinder is very thin, so it is considered two dimensional, which permits visualization of the patterns. Long cylinders in which puzzling alternating stripes of segregated particles can appear have also been studied.

The most common explanation for the patterns noted in some studies is that the angle of repose of different particles can differ, causing differential velocities along the axis of the tube. However, this is still considered an open issue, as particles with very different angles of repose may not segregate under some conditions. (The angle of repose is the angle that a poured heap of particles forms with a horizontal plane and is a characteristic property of a collection of particles.) Under the right conditions, small particles accumulate in the core and then "break through" to the surface, forming bands. If this were to occur in a process, the output of the rotating cylinder would vary in composition.

Theory and experiment have been used to identify some practical consequences. For instance, a non-cylindrical shape, such as an equilateral triangle, may be a better mixer than a cylinder. The degree of fill is a significant variable, with relatively low values performing better. This means that an operator who tries to force larger quantities through a given piece of equipment may be promoting poor mixing.

Another useful finding concerning rotating cylinders is that symmetric baffling makes little difference. Such baffling is common in coating drums. It is better to have just one baffle.

Real mixing situations often involve multiple components, which may differ in size and density. Some may be fragile and subject to breakage. Coatings or other components may be liquids, such as fat or water. Liquids can dramatically change the behavior of solids by forming liquid bridges among particles and changing their cohesiveness. If particles are close in density, reducing size and making all particles similar in size normally improves mixing. However, density and size have opposite effects, so where density may differ, size differences may help to compensate for what otherwise might be a segregating situation.

Cohesiveness refers to the ability of powders to stick together. It can be experienced qualitatively by pressing a small sample between one's fingers. It can be measured in a shear cell, where a sample is compressed and the force needed to slide the two halves of the cell apart is measured.

Cohesive powders can experience flow problems because it is easy for them to form bridges, but they are less inclined toward segregation than non-cohesive powders. Thus, in any specific mixing situation, the geometry, the composition of the mixture, and the potential effect of additives must all be considered. In general, properties of solids are not studied in industry to the same extent that properties of liquids are.

Relatively simple studies, such as measuring angle of repose, or "mixability" in a standard device, could be useful to industrial practitioners. One such device that has been used is a rotating sphere whose axis can also be oscillated. Depending on the frequency and amplitude of oscillation, segregation may or may not occur in binary mixtures. This or another simple system might be used as a standard instrument to study powder mixtures, much as a capillary tube is used to measure viscosity of liquids.

Feeding. Finally, if a proposed mixture must have widely disparate particles, such as croutons in a seasoning mix for stuffing, or the confection called bridge mix, it might be most practical to avoid conventional mixing altogether and just assemble the product in the package. This then becomes a feeding challenge instead of a mixing operation. Feeding is also the critical factor in continuous mixing.

2.1 Continuous Mixing

Continuous mixing occurs when separate streams of ingredients are combined in some device, mixed, and the mixture removed continuously. The device may be a ribbon or paddle mixer in which the feed enters at one end and is withdrawn at the far end or it may be a rotating tube with baffles or V-turns. In any case, the accuracy and reliability of the feeders is the key to success. Some feeders are volumetric while others rely on scales or load cells to continuously weigh a flowing stream. Weigh belt feeders need some positive control over the flow of solids. Some try to control flow by changing the speed of a belt conveyor, but for many solids the range of control by this method is limited. It is better to use the weigh belt as a measurement while running at constant speed and to adjust flow with the speed of a star wheel or screw feeder.

Volumetric feeders can be sufficiently accurate alone when the bulk density of a material is constant, but the bulk density of many solids changes with aeration, compaction, and moisture adsorption. Vibratory feeders are trays fed by slide gates, which control the flow of solids by changing the amplitude and frequency of vibration. Other solids feeders rely on using the angle of repose at rest to stop flow and initiating flow by inducing motion. These rely on the angle of repose being constant, which rarely is true for real materials.

An important consideration for solids feeders is the range of their operation, the ease of cleaning if they are used on more than one material, and their durability in service. The drives and clearances of feeders are usually chosen for specific materials. If the bulk density changes among materials used with the same device, the drive motor may prove inadequate or be oversized. There is a huge difference, for instance, in the density of cheese powder and salt, so that the same feeder does not work well on both materials.

Likewise, the screw design on a screw feeder may be specific for a given density and may not be sufficiently strong if the density increases. Finally, many solids are abrasive and can cause excessive wear if clearances are too tight in screw or star wheel feeders.

2.2 Addition of Liquids

Liquids are added to solids mixtures because they are components of the product, such as fat in dry soup mixes; to control dust by adhering fine particles to larger ones; and to improve uniformity of mixes by agglomerating fine particles of micro ingredients to larger particles. Liquids may be oils, water, or volatile solvents that are later evaporated. In all cases it is important that the liquids be well dispersed. There have been situations where liquids were added by simply pouring from a bucket or pumping from a hose. This is not likely to give uniform mixtures.

It is better to spray liquids from nozzles designed to give a mist of the specific liquid. Liquids can cause solids to build up on walls and agitators of mixers, so it is important that the liquid be added over a length of time and that it be sprayed onto the solids. One proprietary mixer specifically designed to incorporate liquids into solids, such as molasses into animal feeds, uses a slowly rotating cylinder with lifting fixtures that create a curtain of falling solids under a spray nozzle. Other mixers, such as tumbling V-shaped vessels, have a rapidly spinning dispersion bar for adding liquids and breaking up lumps that may form.

2.3 Specific Mixers

It is common to mix many materials in ribbon blenders – half-cylinder vessels with a horizontal shaft around which is wound one or two helical bands or ribbons. Sometimes the agitator consists of paddles on short arms attached to the horizontal shaft. In either case, the objective is to cause convective mixing by moving portions of the mixture around the vessel.

A well-known machine is an enclosed cylinder, which has plows attached to a horizontal shaft. The plows clear material from the side of the cylinder and circulate it back to the center. In addition, choppers – four-bladed agitators inserted between the plow paths, each with its own drive – disperse clumps that may form from liquid addition. Such mixers are used for encapsulation, adding lecithin to dispersible drink mixes, coating particles, granulation (where particles are deliberately grown in size, usually to make material for tableting), and many other purposes, including evaporation and drying.

The vertical shaft, orbiting mixer is a much-lower-shear machine in which an auger extends along the edge of a conical vessel and rotates about its own axis while it circles the vessel. The effect is to gently lift material without causing damage to fragile particles. These machines can be used as mixers or as holding vessels, to reduce or counteract segregation. They can be difficult to empty completely because the exit port is along one side. Also, if the shaft requires a bottom bearing, wear on the bearing can be high because it is immersed in the powder. Smaller machines do not need a bottom bearing.

Ribbon blenders are less expensive than vertical cone screw blenders. Ribbon blenders also use less headroom but have a bigger footprint. The vertical cone screw blender uses less power than a ribbon blender. Both types can have difficulty in completely emptying – the vertical cone because the exit is on the side and the ribbon or paddle because the bottom is level and the agitators must have some clearance from the shell.

Change-can mixers can handle very viscous materials with rotating agitators. The mixtures are held in a removable cylindrical vessel, which can then be transported to filling or further processing. Examples of materials made in such equipment include cosmetics, thick sauces, and high-solids concentrates.

2.4 Examples

1. A mixer with about 50 cubic feet of working volume is used to make a variety of dry powder consumer products that are packaged in pouches containing 8 oz each. The current packaging machine operates at 50 pouches/min. There is a holding bin beneath the mixer that can hold two mixer volumes. The ingredients are supplied in 50 lb bags, but because of partial weighs, i.e., less than 50 lb amounts in the formula, there are about 40 actions required of the operator to assemble a complete formula – dumping a bag or scooping a partial bag. Each action takes about 30 s, so it takes about 20 min to assemble a formula. Mix time is 15 min and it takes about 5 min to dump to the hold bin. The owner wanted to double packaging rate and wondered if mixing can keep up. Bulk density of the average formula is 30 lb/CF.

   Packaging consumes 25 lb/min at 100% efficiency and 22.4 lb/min at 90%. Double is 50 or 45 lb/min. The current system is reasonably in balance but would not keep up with a higher packaging rate. Possible solutions are

   • only increase packaging by 67%,
   • test to see if a shorter mixing time is sufficient (many mix times are arbitrary and may be too long or too short), and
   • add a feed bin to hold a mixer load of ingredients so loading ingredients can occur while the mixer is processing the previous batch.

   Adding a bin was recommended if space was available. Ideally, the bin would be overhead and would be filled through a bag dump station at grade. Head room was tight in this particular case, so the bin would be offset and unloaded through a horizontal screw conveyor.

2. Most commercial operators of dry mixers overfill their equipment. In one case, operators had, on their own initiative, increased recipe quantities until their ribbon mixer was so full that they had to force the lid closed. Not coincidently, the owner had received complaints from customers about inconsistent product – a food flavor. The owner was advised to reduce the formula quantity so that the mixer was about 60% full and the problem with consistency was resolved.

3. Another blender of dry spices also was overfilling and was advised to reduce from 87 to 60% of the volume. This required recalculation of the recipe and adjustment to make assembling convenient. As an exercise, recalculate the formula, given the following data and understanding that final volume can safely range between 55 and 65% of the total. Bulk density of the mix is 37 lb/CF and total volume of the machine is 64 CF.

   What would be some criteria for convenience? One approach would be the number of actions to deliver a given weight, where an action is dumping a complete package of an ingredient or scooping and weighing a portion of a package. Maximizing throughput would be another approach, within the constraints given. Typically mix time does not change significantly with load as long as there is sufficient room for the mix to move.

Ingredient	% Weight	Package size, lb
A	52	Bulk bags
B	27	50
C	7	22
D	5	11
E	4	20
F	3	15
G	2	15

4. A manufacturer of food ingredients used a continuous V-blender to make a wide range of products from about 12 different ingredients with different flow properties using weigh belts to control the composition. The only real control on solids flow was the gate opening on the storage bins. Some solids flooded while others periodically bridged and stopped flow, at which point the weigh belts would speed up fruitlessly. There was also a poor practice of repeatedly changing the calibration on the weigh belts. It was difficult to change the design, but a better approach would have been to replace the simple slide gates with positive feeders linked to the weigh belts. One typically cannot reduce the flow of free-flowing solids, such as soy flour, by reducing the speed of a belt. The flour will simply overflow the belt and its enclosure until the mechanism jams. I learned to be very skeptical of claims to control flow with weigh belts. I believe they are good measuring devices but not control devices.

5. Solids feeding is a complex topic. As previously mentioned, there are positive feeders, including star wheels and screw feeders. There are various gates and valves, which may be adjustable, so that the flow area is varied, but this can be difficult in some solids, so gates and slide valves are normally opened all the way or to a predetermined position, and not used to control flow. Often a gate from a bin is combined with some form of flow stimulation, including vibrators on the bin side wall and air injection through nozzles in the walls.

Solid powders often need stimulation to flow – when at rest they stay at rest. Advantage is taken of this property in a flow control device that looks like a Venetian blind. The angle of the slats is chosen so that the solids in the bin will not normally flow through the gaps because of the angle of repose. The device is vibrated when flow is desired, and the frequency of vibration can affect the rate of flow, because the angle of repose changes under vibration. One advantage of this device is that it does not require a horizontal offset, as do screws or vibratory trays.

Loss-in-weight feeders are bins with controllable discharge devices mounted on load cells so that their weight is continuously measured and the discharge device modulated to achieve a desired flow rate. The bin on a loss-in-weight feeder is usually fairly small to maximize accuracy and minimize cost. This means that it must be refilled, but during filling, the weighing function does not work. The discharge device runs at the last rate it was using, referred to as volumetric feeding.

Volumetric feeding can be accurate for solids with constant bulk density, but many solids change density under vibration, pressure, or aeration. Thus, the usual objective in designing a feeder is to minimize the time during which a loss-in-weight or gravimetric feeder is in volumetric mode. A target of 10–20% is common. This requires sizing of the feeder bin as well as the supply bin in relation to the target flow rate of solids, and the selection of a discharge device for the supply bin, which is normally installed above the feeder. The gravimetric feeder has instruments to measure high and low levels, calling for refill when the level is low and stopping refill when the high level is reached. The supply bin may be replenished by pneumatic lines or by dumping bags either directly into the bin or through a ground level bag dump station. The low level in the feeder bin must maintain a "heel" of solids so that flowing material does not just flow right through the open slats. To enable start-up, such feeders often are fitted with gates that close so that the heel can build up.

A system as described was created for a pet food plant in which fairly complex mixes were fed ultimately to a cooking extruder. The supply bins were arranged in a circle within a work platform. Each had a loss-in-weight feeder beneath, and these fed chutes to a bin for collection of the formula. As an exercise, specify the sizes of the supply bins and gravimetric feeders, given the following information:

Ingredient	% Weight	Bulk density, lb/CF
A	50	50
B	30	60
C	11	40
D	5	20
E	3	25
F	1	30

The desired total flow rate is 5,000 lb/h. (The formula and densities are arbitrary and do not necessarily represent a real product, but are used here for illustration.) To proceed, calculate volumetric flow rates and then pick working volumes of the feeders that will last for some length of time. (What should that time be?) Assume some replenishment flow rate. (How would you do that? The answer actually requires knowledge of the solids properties, which have not been provided. Some may aerate easily, some may be dusty or abrasive. For this exercise, assume a refill flow rate of 50 CF/min by gravity flow through a slide gate.) (Instructors can change the given parameters to get new versions of the exercise.)

One approach to this type of system is to fill the supply bins with a day's or a shift's requirements, hence the common name of "day bin." How big would they be if this were the approach? Another approach is to reduce capital by making the supply bins smaller and replenishing them either automatically or manually during the run. Consider the trade-off if labor costs $15/h, one person can tend the system, and bins cost about $5,000/m³ with a scale factor of 0.45 (Maroulis and Saravacos 2008, p. 108). The scale factor is the exponent in the equation

$$C_2 = C_1 \left(V_2/V_1\right)^n \tag{2.2}$$

where

C_2 is cost of second vessel, \$
C_1 is cost of first vessel, \$ (use \$15,000 in this case)
V_2 is volume of second vessel, m^3
V_1 is volume of first vessel, m^3 (use 20 m^3).

The costs of many types of process equipment correlate with an equation of this form and often have exponents around 0.7. Corrections should be made for materials of construction and for inflation, which are being ignored here.

2.5 Some Lessons

1. In dry mixing, getting ingredients into the mixer and removing the mix in a timely manner can be as important as having the right mixer and mix time.
2. Provision of staging bins and surge bins can increase the productivity of a given mixer in a system, at the cost of additional equipment and space.
3. Most dry mixers are probably overfilled, so consistency can often be improved by reducing the batch size.
4. Optimum mix time is rarely determined for each formula. Rather it is usually arbitrarily chosen and is likely to be too long or too short. Determining mix time is tedious but worth doing.
5. Formulas can usually be modified to maximize the use of unit quantities of ingredients – whole boxes, drums, or bags.
6. Continuous mixing is a feeding problem, not a mixing problem.
7. Feeding solids has its own challenges, including measurement, flow control, wear from abrasion, and cleaning of complex equipment.

Chapter 3
Snacks and Baking

Cereal-based foods are a significant segment of the food industry. Most of them have in common the process steps of mixing, forming, baking or frying, finishing somehow (such as by icing, seasoning, slicing), and packaging. Some products are fermented while others are not. Different raw materials require different means of handling, cleaning, size reduction, and conditioning.

Popular snacks are made from potatoes, corn, nuts, sugar, dairy products, and wheat by a variety of processes.

3.1 Snacks

One of the most popular salty snacks, the potato chip, was 150 years old in 2003. George Crum, a cook at Moon's Lake, invented the potato chip at Moon's Lake House in Saratoga Springs, NY, in 1853. It quickly became popular in restaurants along the East Coast. Later innovations were packaging in small bags by Laura Scudder in Monterey Park, CA, and larger-scale production by names still familiar such as *Wise, Utz, Mike-sells, and Lay's*. Today, retail sales in the United States are about $6 billion/year.

Corn-based snacks are increasingly popular in the United States, including tortilla chips, corn chips, and corn puffs. Rice-based snacks are popular in Asia. These products, whether based on corn, rice, or mixtures or grains, are formed by extrusion, sheeting, or pelletizing and then puffing.

There are differences in flavors between snack foods in other countries and those popular in the United States For example, seaweed and shrimp flavors are popular in the Orient but are rarely found here. Dairy-based flavors, such as sour cream, are common in Scandinavia, and almost anything with paprika sells in Germany.

Other trends are manipulation of fat content in fried snacks by using higher temperatures, baking instead of frying, and steam stripping to remove excess oil from fried snacks.

J.P. Clark, *Case Studies in Food Engineering*, Food Engineering Series,
DOI 10.1007/978-1-4419-0420-1_3, © Springer Science+Business Media, LLC 2009

3.1.1 Technical Platforms for Snacks

Frying is still one of the most significant unit operations or technologies, and it has many variations. Frying can be continuous or batch; take place under vacuum, atmospheric pressure, or elevated pressure; and may involve immersion of food pieces in hot oil or spraying with hot oil.

Normally the oil used is vegetable oil, but the oil can also be olestra, a sucrose polyester that is less digestible and thus less likely to contribute unwanted calories. In some fried snacks, oil may be 30% of the final product, contributing to both flavor and caloric content.

Motivated by health concerns of consumers, manufacturers have developed snacks with less fat. One technique is steam stripping to remove some oil while retaining the characteristic flavor and texture of traditional potato and corn snacks.

Other approaches are baking or drying of "fabricated" snacks made from dehydrated potato and baking instead of frying extruded corn curls. These have no added oil and usually rely on added flavors. Application of flavor to baked potato and corn snacks can be challenging because the chips are fragile and lack the oil that in fried snacks assists adherence of powdered seasonings.

Other baked snacks include pretzels, which are based on wheat dough and formed into various sizes and shapes. Pretzels acquire their characteristic surface sheen by a dip in a baking soda or lye solution. This same sheen makes adding flavors a technical challenge. There are proprietary technologies for applying flavor systems to coat pretzels, including the application of a water solution of starch to act as a "glue." This then requires a drying step achieved either in an additional piece of equipment or by passing warm air through the coating equipment.

Another major technique to make snacks is extrusion, which may be high shear or low shear. For example, a high-shear, relatively short extruder is used for puffed corn snacks and more-dense corn chips. Low-shear extruders are used for pasta and for unpuffed pellets. Pellets are sometimes shipped from a central plant to locations close to markets where they are fried to a final form for sale. In other cases, the pellets are used almost immediately in a complex process to produce three-dimensional shapes.

Still other snacks are made by sheeting and cutting. Examples are tortilla chips made from ground whole corn, which has been soaked in lime slurry to soften the skins. After sheeting, tortilla chips are generally fried but can also be baked.

Washing the soaked corn removes the skins and creates a wastewater stream. It is possible to conserve water in a corn snack plant by reusing water first to wash the soaked kernels and then as part of the soaking process. Suspended solids are removed by filtration.

3.1.2 Equipment Innovations

Twin-screw extrusion is often used for higher volume production of low density, i.e., highly puffed snacks, such as corn curls. New variations include snacks with two colors in the same piece and filled salty snacks, achieved by co-extrusion of filling and exterior doughs.

Extruders' screw diameters range from 32 to 187 mm, and are designed to make scale-up more reliable. Surface-to-volume ratio is not constant as machines become larger, so some manufacturers have developed proprietary software to model heat transfer in twin-screw extruders. This enables accurate prediction of performance based on experiments at a small scale. A useful scale-up parameter is specific mechanical energy, which is energy per unit mass flow (Valentas et al. 1991, p. 274).

Low-shear extruders and drying systems are used for pasta and to manufacture pellets, as mentioned above. Pellets enable high-efficiency production at central locations and completion of the process close to market. The final products usually have very low bulk density and can be fragile, so shipping them long distances is expensive.

3.1.3 New Frying Technology

Most fryers immerse products in a bath of oil heated either by coils in the bath through which steam or hot gases flow, or heated externally in a heat exchanger. Products are moved through the oil bath with a chain or belt conveyor, depending on whether the products sink or float in the hot oil. Some products, such as donuts, which float, must be turned over part way through the bath to get even cooking. Most products pick up oil as moisture is driven off. This removal of oil is compensated by adding fresh oil to keep the bath level constant. Depending on the volume of the bath and the rate at which oil is removed with product, oil may be in the bath, constantly being heated and exposed to oxygen, for varying lengths of time. Some rancidity is almost inevitable.

A new type of fryer uses falling curtains of hot oil to enrobe food products passing beneath them on a wire mesh belt. The complete fryer system operates with a very small volume of oil. As a result, oil carried out of the fryer with product is quickly replaced with fresh oil, resulting in a rapid oil turnover rate, so foods taste fresh and have a long shelf life. This concept was developed especially to minimize oil quantity and maximize removal of fines. In standard fryers, product particles remain suspended in the oil and are carried along with it, causing off-flavors to develop and promoting degradation of the oil. The new fryer uses the returning product belt to drag along the oil collection pan and move any fines to the in feed of the fryer, where they are discharged into a continuous filter for removal.

Most modern fryers have external heaters and filters for the oil to provide good temperature control and fines removal. Overflow weirs above the moving belt create the falling curtains. By controlling the oil flow to each weir, the heat transfer rate can be modified as product moves through the machine. The same fryer can be used with hot water as a blancher. Deep beds of nuts have been roasted with the falling curtain.

One paradox of snack processing that is illustrated by several of the examples described here is that new products, especially those catering to health-conscious consumers, rely on more complex processes and equipment, such as steam stripping, extrusion, and new types of fryers. On the other hand, these same technologies can find wide application in other areas of food processing, as exemplified by extrusion of pet foods, breakfast cereals, and fish foods and by frying of shrimp and roasting of nuts.

3.1.4 Coating and Seasoning

Coating and seasoning have different purposes and methods for various products. The equipment and processes are unique and are interesting to understand. Coating and seasoning are used to add flavors to the surface of pieces. This may be compared with adding flavor to the base cake or substrate. Often, flavors added to the dough used to make a base cake or other substrates are delicate and are damaged by the frying or baking process. Flavors on the surface are encountered first and so have a greater impact than flavors within the substrate.

Dry seasoning. Dry flavors and seasoning such as salt, powdered cheese, and dry onion or garlic may not adhere well to some snacks, while other snacks, with a residual layer of frying fat may pick up and retain plenty of applied powder. Shiny snack pieces such as pretzels are especially challenging because the sodium carbonate or bicarbonate bath in which they are dipped before baking seals the surface and creates the characteristic surface finish. For these and other baked pieces, it is customary to spray a light coat of oil first and then apply the powder.

There are many ways to apply oil and powders, but one common method is to use a rotating metal drum with internal lifters. Drums come in diameters of 28–60 in. and lengths of 3–18 ft. Typically, the drums rotate at about 10 rpm and are filled so that there is a bed from about 6 o'clock to 9 o'clock as one looks at the end of the drum. Pieces are introduced at one end and exit from the other. Flow is controlled by the rate of feed, the rotation speed, and the elevation of the feed end of the drum.

It can be a challenge to measure the residence time of pieces, but one way is to spray paint some pieces, drop them into the feed, and retrieve them from the exit, measuring the time when they appear. Depending on the piece shape, weight, and flow through the drum, there is some back mixing, so there is a residence time distribution. Typically, the target is about 30 s average when just applying oil or other liquid and 60 s when applying both oil and solids.

Oil is applied through a spray bar with up to six piston nozzles on which the stroke and frequency of pulsing can be adjusted. Oil or other liquid flow is tested by capturing in a cup for a short length of time. A calibration curve can be prepared for a given system and given liquid. Sometimes the oil is heated while at other times it is not. It is usually desired that the oil be relatively viscous so that it adheres to the piece. Typical oil applications are 2–20% by weight. The higher values are used for crackers.

Dry seasoning is applied with a screw feeder inserted at the exit end of the drum. It is adjusted by varying the screw speed and by blocking some of the holes in the feeder tube. To calibrate, seasoning is caught on a tray for a short time period and weighed. Target dry application is 5–12% by weight. Seasonings vary widely in bulk density, so feeder tube diameter and even drive motor power must be adapted to the material. Straight salt is much more dense than dry cheese. Cheese and other seasonings may be cohesive, meaning they stick together well, which can inhibit flow from the feed hopper. It is not unusual for seasoning feed to be interrupted and some product to escape unseasoned.

It is one thing to set the flows in the anticipated ratios; it may be quite another to confirm that the desired final composition is achieved. Some pieces are so uniform that the difference in weight between uncoated and coated pieces can be measured by weighing the same number of pieces from the feed and exit streams. For other, more variable, pieces, it may be necessary to measure a tracer like salt or oil by chemical analysis.

An alternative to oil to achieve adherence is an aqueous starch solution, which acts like a glue for the dry seasoning. However, this approach then requires removing the water using heated air. This could be done in a separate dryer or in a special drum with perforated walls and a plenum or second shell so that hot air can be passed through the bed. The equipment to dry the pieces adds an extra cost and the additional handling if using a separate dryer could increase breakage.

There are other ways to apply oil and dry seasoning, including simply spraying the products on a flat belt as they exit an oven and sprinkling salt or seasoning from a feeder over a belt, but these only treat one side of the product. However, that is adequate in some cases.

Coating with liquids that create a solid shell. Some products are coated by dipping in chocolate, such as ice cream bars, while others are run through a falling stream on a wire mesh belt in a process called enrobing. Coating with a melt that solidifies as it cools is a complex situation to model. There is gravity-induced flow of a fluid with a changing viscosity, heat transfer that changes the viscosity, and surface adhesion. In practice, experiments are performed to establish the correct temperature of application, the optimum time of transit beneath the waterfall, and the time for cooling. Excess coating is recovered and recycled, so it is desirable to minimize inventory by keeping the reservoir small and it is critical to control temperature carefully, especially with chocolate, because chocolate is sensitive to temperature and can solidify in unstable crystal forms if it is not properly handled. These forms can create cosmetic defects in coated products.

Panning. A very wide range of products is made by coating centers with sugar or chocolate in a process called panning. Centers may include nuts, fruits such as raisins or cranberries, soft gels as in jelly beans, hard candy, chocolate lentil-shaped pellets and, in the case of pharmaceuticals, tablets of drugs. The common feature is that the centers acquire a coating from successive applications of syrups or melted chocolate, which are transformed into a solid shell by various means. The pieces polish and shape each other by tumbling together in a rotating vessel. There is a great deal of art in the specific operations, which are often done in batches and still involve manual labor and close attention.

The conversion of a liquid to a solid shell is a phase transformation that may be assisted by temperature and humidity control. In the case of chocolate, the transition is from a melt to a crystallized solid fat using cold air. In the case of shells over soft centers, a sugar syrup that also contains color and flavor is dried with dehumidified air and the addition of dry solid sugar. In the case of a sugar shell over harder centers, sugar syrup is dried using warm air. Normally, the centers are loaded by hand, syrups and solid sugar are spread by hand, and finished pieces are unloaded by hand. One person can tend about four pans at a time in many cases.

Reducing the labor involved while achieving greater consistency is one objective of automated pans. Automated panning systems load, coat, and unload pieces using conveyors and a programmed controller. Sugar coating is often about 30% of the piece final weight while chocolate is often 50%. Batch times for chocolate are 2–3 h. The automated equipment permits coating and polishing in the same unit, while in batch systems, there is a transfer to another pan.

The advantage of the automated machine is its use of a very shallow bed depth, which protects soft centers, reduces batch times, and enhances uniform coating. The equipment is more expensive than a comparable set of conventional pans and is suited to relatively large batches of 500–3,000 kg as compared with about 200 kg in a typical single pan. Automation can also be flexible, in contrast to intuition that might suggest it to be best for running the same product a long time. The typical machine has a volume of about 1,300 l, and the capacity obviously depends on the bulk density of the product.

The vessel is a perforated horizontal cylinder with a shroud or plenum around it to provide air flow, in contrast to the duct or pipe pointing into the mouth of a conventional pan. Each of the many steps in whatever process is performed is programmed and sequenced, based on the specific product. The ability to reproduce the steps exactly is one advantage of automation. As one extreme, the manufacture of jawbreakers, which are 100% sugar, starts with literally a grain of sugar and builds layer upon layer over a period of perhaps a week.

Conventional pans come in various sizes, 32–60 in. in diameter. Speeds are 22–28 rpm. Some are round while others are tulip shaped – having a slightly straight side. Once coated, panned products are transferred to polishing pans where a small amount of carnauba wax or confectioners glaze is applied to protect the outer surface. Polishing pans have ribs on the inside surface to help mixing. The flow patterns of pieces and syrup in the different sizes and shapes vary. The pans are about 30% full by volume after coating is finished. This is as full as they can be without pieces falling out. Some may attempt to increase the angle of the pan to increase the weight per batch; however, there is an optimum angle that ensures constant movement and constant coating in the pan. The product can also limit pan loads. Lower loads are used for soft centers, which could deform under their own weight. Various flavors of jelly bean coatings behave sufficiently differently that a given operator tends to specialize in one or two colors.

3.2 Baking

Baked goods include bread, cake, cookies, crackers, and flat breads, such as tortillas. Many are leavened, meaning they are foams in which the bubbles are created by carbon dioxide released by fermenting yeast or by the chemical action of baking soda (sodium bicarbonate). Each step in the manufacture of a baked product contributes to the characteristics of the product (Blanshard et al. 1986, Matz 1988, among many others).

Most baked goods are based on wheat flour, though other cereals are also used, alone or in mixtures. Wheat contains a unique protein, gluten, which has the ability to form a strong network when it is hydrated. This network stabilizes the foam that is formed in leavening. The other major component of flour is starch, which gelatinizes in the presence of heat and moisture. Gelatinization refers to starch losing its crystalline structure and becoming soluble and amorphous. Starch is a chemical polymer composed mostly of glucose monomers, though other sugars and carbohydrates also occur. Starch can be hydrolyzed after it gelatinizes, meaning that it is broken down to smaller molecules and sugars. Yeasts feed upon the sugars and produce carbon dioxide and ethanol. Depending on the desired final product, the key steps in making baked products are mixing, forming, baking, and finishing.

3.2.1 Bread

Bread can be made from highly purified white flour, whole wheat flour, mixtures of the two, and mixtures with other cereal flours, such as rye, barley, and oats. When bread contains high fiber flour, such as whole wheat or flours of other cereals, a higher gluten content is required in order to maintain the desired texture. If the wheat flour does not have enough natural gluten, then purified gluten isolated from other supplies of wheat is added. The operation of isolating and purifying gluten is an interesting process in its own right.

Gluten. Gluten is isolated from wheat flour by washing with water to remove the starch, leaving gluten as a gummy mass. The starch slurry is concentrated by settling or by centrifuge and the starch is dried as a product. The gluten is also dried and ground to a powder. Wheat starch is used industrially for textile and paper sizing and for making pastes, primarily where less expensive corn starch is not available. (Sizing refers to the addition of a filler to paper and textiles to improve the surface finish for printing.) Gluten isolation also uses large quantities of water, so the process typically occurs in countries other than the US, especially Australia and China. Thus most gluten for food use as an additive is imported. For it to be useful in baking, it must retain its gel-forming functionality, which requires careful, low temperature drying, since gluten can be denatured by heat because it is a protein. Denatured gluten still has nutritional value and is sometimes added to foods for that purpose.

Returning to bread making, the basic ingredients of bread are flour, water, salt, and yeast. Typically it requires about 30% moisture to form a proper dough, which is a coherent mass that is plastic and not sticky at the surface. The act of mixing water into the flour becomes kneading, which means working or massaging the mixture to form the strong network of hydrated gluten. This consumes considerable energy and requires a sturdy machine. Typical dough mixers are horizontal vessels with one or two shafts shaped into short blades. It is common to monitor the energy consumption of the agitator drive as a measure of the development of the dough. The objective is to cease kneading when energy consumption is near its peak, because continued mixing has been observed to breakdown the dough structure and produce a weak bread structure. One exception is English muffins, which are deliberately overmixed to produce their characteristic open cell structure.

In addition to the basic ingredients, some breads have sugar, shortening, and other ingredients added. Some ingredients are intended to correct for variations in the flour by acting on the gluten to make it stronger or weaker by oxidizing or reducing chemical bonds in the protein. Other ingredients are intended to extend shelf life.

Shelf life of bread. The shelf life of bread is normally related to its texture. Bread is expected to be relatively soft, but with time, it hardens. This is called staling. It seems that it is drying out, but in fact there is very little moisture loss. The change in texture results primarily from recrystallization of the starch, known as retrogradation. Additives that inhibit retrogradation extend the time period during which bread remains soft. Some such additives are emulsifiers, such as mono- and di-glycerides, chemicals made by reacting purified fatty acids with glycerol. Other effective additives are enzymes that affect starch's ability to crystallize. The final determinant of shelf life is mold formation, which can occur if the surface gets wet, as by condensation of moisture in the package. To extend shelf life, yeast and mold inhibitors may be added, such as sodium propionate. Vinegar is less effective but may be preferred as a more natural ingredient.

Additional added ingredients include malt, yeast food, and other enzymes that accelerate the production of sugar for the yeast to consume. Yeast food includes nutrients that may be in short supply in wheat flour. For some products, other ingredients include whole grains, caraway seeds, caramel color, and other sweeteners (molasses, for instance).

Bulk ingredients, such as flour, oil, and sugar, are typically delivered to the bakery in bulk trucks, unloaded with pneumatic systems (or pumps, for liquids), stored in silos, and delivered to the mixer by pneumatic lines. Amounts are measured by having receivers on load cells. The mixer may also be on load cells, to confirm that a mix total is correct, but because of its weight and the vibration during mixing, a mixer is not a very accurate measuring device. Minor and micro ingredients are added by hand from bags, drums, or containers holding pre-mixes that are prepared off-line.

After all ingredients are added and the dough is properly developed, it is dumped from the mixer by tilting the vessel and partially filling a rolling tub called a trough. Troughs are sturdy and hold a typical mixer load of 2–3,000 lb. In the typical process, troughs are rolled into a temperature and humidity controlled room, where they may remain for 2–24 h. During this time, the yeast consumes sugars and generates carbon dioxide, causing the dough to roughly double in volume. For some products, the dough may be "punched down," meaning that the gas is partially released and the dough collapsed, only to rise again.

Efforts have been made to ferment continuously either in a tower filled with dough or by putting troughs on a chain that moves them slowly through a room. The tower proved problematic when there were process interruptions, causing dough to overflow and to expand more than intended. The moving chain guarantees first-in-first-out management of the dough but, again, is less flexible than the system in which each trough is independent. Baking is still a hands-on and somewhat artisanal business in which experienced bakers are sensitive to subtle variations. For making good quality bread, bakers prefer to have flexibility.

Some of the continuing challenges of baking are to shorten the time for production and to reduce the manual labor involved. The traditional process previously described is known as the straight dough process. Some variations intended to improve labor productivity include the sponge and dough process and pre-ferment or brew process.

Sponge and dough process. The sponge and dough process prepares a thin dough or flour and water mixture, which may be capable of being pumped. After it is allowed to ferment in a vessel, it is returned to a mixer where additional flour is added to make a finished dough. This dough may rest briefly before continuing through the process, but is not conventionally proofed.

Brew or pre-ferment process. In the brew or pre-ferment process, an even thinner slurry of flour, water, and other ingredients is prepared and allowed to ferment in a tank. It is combined with flour, often in special continuous mixers, which also aerate the dough and is pumped to subsequent steps. Both of these variations use fermentation to generate some traditional flavor and leavening, but do so with only a portion of the total flour to reduce the time involved and to have a material that is easier to transport than traditional dough. One consequence is that, to many people, the flavor is inferior to bread made more traditionally. Also, the faster processes are best suited to plain white bread. Bread with additional grains, particulates, and high fiber content typically are best made in the straight dough process. Buns and rolls, such as hot dog and hamburger buns for quick serve restaurants, are often made with the accelerated processes because flavor is less important than low cost for such products. It is one of the great lessons from bread baking that there is a tradeoff between the time involved and flavor – longer time yields better taste (and higher cost).

Forming. Dough is elastic and retains stress forces when it is deformed. During proofing, it "relaxes" while it ferments and expands. Even minimally fermented doughs need floor time to remove residual stress and permit forming. Forming is initially by removing chunks of dough and then rounding these into a ball or cylinder. Typically, rounding is volumetric in spite of the fact that the density of the dough is continuously changing, because the yeast is still active. As a result, dough balls are usually inconsistent in weight, meaning that the resulting loaves are also variable in weight. Because bread is sold by weight, rounding equipment is set so that the lightest loaf is still above the label weight. This means that bakeries always "give away" bread by selling loaves that are heavier than the label claims.

It is possible to weigh every dough ball on line, using a checkweigher, and adjust the rounder to maintain more consistent weight. This requires fitting the rounder with servomotors to adjust the portioning device precisely.

The dough balls for pan bread are dropped into metal pans, which typically are strapped together in sets of six pans. Dough balls for hearth breads, which are baked directly on an oven belt, are dropped into carriers on a chain. In either case, the dough balls pass through a second proofer for a relatively brief time to allow the dough to relax again. The proofer may be enclosed or, often, is just a serpentine path in the air on the way to the oven.

Baking. Modern baking ovens are continuous tunnels with chains carrying straps of pans or belts on which bread is directly baked. There may be several separately controlled zones, and heat is normally supplied by burning natural gas and circulating the heated air and combustion products with fans. A portion of the air is exhausted while some is recirculated. Baking is one of the more challenging phenomena to model because the geometry as well as the physical properties is constantly changing. When the dough balls are first heated, they expand rapidly in volume as the gas contained in the starch foam expands. This is called oven spring. Water in the dough is vaporized and escapes, contributing to the volume expansion and drying out the dough, which serves to fix the texture of the crumb, or body of the loaf. The surface of the loaf dries first, so the temperature of the surface increases well above 212°F, while the center of the loaf may remain somewhat below 212°F. The wet dough is a better thermal conductor than the drier foam near the surface, so there is a steep temperature gradient near the surface and a flatter temperature profile in the center of the loaf.

Baking and drying is further complicated by the presence of the pan, which restricts moisture removal from the sides and bottom of the loaf. Some bread is baked in perforated pans or baskets, but most are solid. Heat transfer is by convection to the top and conduction from the metal pan to the sides and bottom. Improving external heat transfer, as by increasing the velocity of circulating air, can have only limited impact because most of the resistance to heat transfer is internal to the loaf. One technique that was shown to have some effect uses a bank of needles that are electrically charged to direct a stream of ionized air molecules directly at the surface of the bread. This stream breaks up the stagnant boundary layer near the surface and increases heat transfer rates. It was particularly useful for loaves baked in baskets because it seemed to affect most of the exposed surface and shortened bake time. Another approach uses high velocity air streams directed at the surface. This is called impingement and has been applied to pizza ovens, where improving surface heat transfer is especially relevant.

A major improvement in baking is achieved simply by improving the uniformity of temperature. Altomare (1994) found very wide variation in temperature profiles across baking ovens and through zones. He used a recording array of thermistors that stored data in a solid state device and measured temperature profiles in about

50 different ovens. The devices are commercially available and their use would probably benefit every commercial baker.

Typical bread has about 18% moisture after baking. Overbaking runs the risk of selling underweight loaves. Underbaking produces bread with too soft a texture. Color, crust and crumb texture, and flavor are all affected by baking time and temperature. Crust texture is often modified by injecting steam into the oven, which helps make a tougher crust, as in French bread. The increased humidity prevents the crust from drying out too quickly.

After baking, loaves are cooled, often by simply traveling on a conveyor above the oven after being removed from the pans or, in smaller bakeries, by standing in racks for about 1–3 h. Loaves are at a temperature of about 200°F when leaving the oven. The objective is to reduce the temperature to about 95–100°F, at which condition they are easy to slice. An improvement is to cool in a controlled environment, using conditioned air in about 75 min. Air is conditioned to 75°F and 80–85% relative humidity for optimal cooling. Intake air should be filtered to prevent contamination.

Changing pans on a multi product baking line is a time consuming and tedious project when done manually because there may be several hundred straps of pans on a typical line. Pans not in use must be stored somewhere. Pan handling in a bread bakery was one of the first successful applications of robots in the food industry.

Slicing and packaging. Bread is usually sliced automatically by passing through an array of saw blades and then packaged in plastic bags closed with a twist tie. Maintenance of the slicers is important because broken saw blades are a source of metal contamination. Bags typically are supplied on wickets or stacks of preformed bags, which are placed on machines, which use a burst of compressed air to open the bag, into which the loaf is pushed. Packaged bread is typically placed by hand on plastic trays and the trays stacked on rolling carts. Bread is fragile and needs to be handled carefully to avoid damage. Because of its relatively short shelf life, bread is delivered daily and a product near the end of its code (designated shelf life) is removed from the store and returned to the bakery.

What to do with returned product has been a challenge. It is usually edible and acceptable in texture because code date is deliberately made shorter than true shelf life, so that product sold near the end of its code still has a few days of use left. Large commercial bakeries maintain thrift stores in which returned products are sold at a discount. These stores are popular with people on tight budgets, but care must be taken in locating them so that they do not directly compete with regular customers, such as chain stores. Further, the supply of product to the thrift store is inconsistent, so sometimes perfectly good product is diverted just to keep the shelves stocked. This is not profitable. One motivation for extending the shelf life of bread is to eliminate returns, if possible, and, consequently, thrift stores. On the other hand, thrift stores are profit centers in some corporations and their elimination could be difficult. Faced with the possibility of bread that would last 30 days, bakeries were uncertain how best to realize the benefits, if any.

3.2.2 Cake

Cake is distinguished from bread in that it is chemically leavened, uses a weaker flour (lower in gluten), and usually contains more sugar, shortening, and other ingredients, such as eggs, milk, and flavors. Cakes are mixed with higher moisture than bread and the result is a batter rather than a developed dough. The final product also usually has higher moisture content than does bread. Cakes can range from relatively simple single serve pieces to elaborate multi-layer and highly decorated creations. Icings, frostings, and fillings add different textures and flavors. Most are high in sugar and fat.

Because batter is thinner than dough, most cakes are baked in pans. Both bread and cake pans are sprayed with a release agent, often mineral oil, but cake pans are washed after every use, while bread pans usually are not.

If a cake is to have a coating or filling, it must be prepared in its own mixer. This is typically done close to the point of application, which means ingredients must be brought to that spot, packaging material removed, and the equipment cleaned after use. These requirements affect plant layout and operation. Fillings, icing, and coatings are often sticky and easily get on equipment and floors, meaning a cake bakery has greater sanitation challenges than does a bread bakery.

Cakes usually have relatively short shelf lives, comparable to those of bread, when made with relatively high moisture. Shelf life is normally limited by staling or hardening of the starch foam, but may be affected by other phenomena, such as breakdown of icing or coating. Moisture migration from the crumb to icing can cause the icing to dissolve and disappear. Icing is an opaque slurry of very small sugar crystals created when a supersaturated solution of sugar cools rapidly. The water activity of icing is lower than that of cake crumb, so moisture moves from the cake to the icing. If it is possible to make a higher moisture icing, this migration can be reduced. Most materials tested as potential barriers to moisture migration have not been effective, though some work in other applications. Candidate barrier materials include mono- and di-glycerides, whey protein films, and other protein films.

Doughnuts are sweet goods that are similar to bread and cake but are prepared by frying. Yeast-raised doughnuts have recipes similar to those of bread and stale very quickly after frying. Cake doughnuts start with batter and chemical leavening and have longer shelf lives than yeast-raised doughnuts. Some doughnuts are injected with jelly or cream filling and some have icing or frosting applied. One approach to extending the shelf life of yeast-raised doughnuts recognized that they did not have to be perfect right after frying but rather a few days later, when they would be consumed. This led to modifying the frying operation so that the doughnuts improved with age before beginning to stale. Balancing the water activity of the doughnut and the icing also extended the shelf life before the icing broke down.

Some snack cakes are formulated with lower moisture to extend their shelf life. These typically have higher sugar content than do moister cakes and because they are dry to start, any starch retrogradation is not noticeable. Such snacks often have compound coatings and low water activity fillings, such as jams and jellies. While

moist cakes are delivered frequently and have code dates of 5–7 days, lower moisture snacks may have no code at all and are not removed from the store shelf until
they sell.

3.2.3 Cookies and Crackers

Cookies and crackers are dryer baked goods that are usually relatively small, individual pieces. They are mixed as rather dry doughs or batters and are baked quickly
at high temperatures in very long ovens. Bake times may be 2–3 min compared to
18–20 min for cake and 20–30 min for bread. Cookie and cracker ovens may be
300 ft long and 36–42 in. wide. Belts are solid stainless steel, perforated or woven
wire mesh.

Forming. Pieces are formed by sheeting (rolling out) ribbons of dough and cutting
or stamping out pieces, by pressing dough into molds machined into rolls, by cutting off pieces from an extruded cylinder of dough, or by co-extruding dough around
a filling. Each approach gives somewhat unique results. Sheeting, for instance, permits laminating, in which thin layers of dough are folded over each other, sometimes
with a coating of oil or butter in between. After each fold, the ribbon is rolled again
and the procedure is repeated at least three times. The result is a flaky textured characteristic of saltines and other savory crackers. Previous to sheeting, the doughs for
crackers are usually allowed to ferment in a temperature and humidity controlled
room. The dough is inoculated with a mixture of yeast and bacteria. Before the
understanding of the micro culture responsible for the taste and texture of crackers,
the process involved simply never cleaning the troughs in which the dough was held.
The residual dough scraps and the scratches and crevices of the steel tubs maintained
an adapted culture that lightly leavened and acidified the dough.

Crackers puff a bit upon baking, but since many are packed in slugs or tight
stacks, it is important to rigidly control their thickness. One control point is the final
sheeting, but the degree of leavening and the temperature profile of the oven also
matter.

Molding permits the precise formation of one surface and good control over piece
weight, although dough density can vary. Butter cookies, sandwich cookies, and
others with a distinctive surface pattern are made by molding.

Wire-cut or drop cookies are closer to the way many homemade cookies are
formed. A cylinder of dough is forced through a nozzle and a wire is drawn through
to remove a portion, which falls on the baking belt. During heating, the dough
"melts" and flows so that a mostly uniform disk is formed. Piece weight and shape is
more difficult to control, but certain products are expected to have the rough surface
and variation that wire cutting gives. Some products that look like they are formed
by wire cutting are actually formed by molding, using molds that are shaped to look
random.

Co-extrusion forces a stream of filling through a center nozzle and a circle of
dough through another nozzle around the filling. The result is a tube of dough
with filling inside. A popular filling is a fruit jam made using figs, strawberries,

or other fruits. The tubes pass through the oven. In all cases where a single layer is placed on a belt, pieces are located as close together as possible while leaving room for steam to escape. The tubes are cut into short pieces after baking and the pieces are shingled together and stood on one end before entering the cooler, which is a refrigerated tunnel. Fruit-filled cookies are heavier and have a higher heat capacity than plain cookies and so require controlled cooling with refrigerated air. Because the jam is sticky, cutting can be a challenge. One solution is a water knife, which uses a very thin and very high pressure stream of water to cut. The pressure may be 50,000 psi, achieved by using a pressure intensifier, similar to a hydraulic pressure system, in which a large piston drives a much smaller one.

Packaging. Cookies and crackers are often packaged in slugs or stacks wrapped like coins. This requires that pieces be very uniform because the packaging equipment simply grabs all the pieces within a certain length. If some pieces are too thick, they get crushed. Other products are placed in small stacks, in rows in a box, or randomly dropped into bags or pouches. Some packaging is automated, but much is still done manually. The work can be repetitive and has caused injuries, leading to substantial fines against employers. Light weight, high speed pick, and place robotic arms have been applied to cookie and cracker packaging. These use vision systems to locate pieces on a belt, pick them up using a vacuum tool, and place them in stacks in a box. An advantage of using manual labor for packaging is the opportunity for a final inspection for color, piece integrity, and general appearance. Manual packing is also more flexible and requires lower capital, so is often used for new products whose ultimate success is undecided. An example is mini versions of popular sandwich cookies, which proved to be a short-lived fad. Packing was manual, involving scores of workers, but the equipment cost was minor, though the space occupied was substantial. Specialized packaging equipment would have had a short useful life and cost millions of dollars.

Packaging in cookies and cracker bakeries occupies much more area than does mixing and baking so a typical layout may involve two or more floors for packaging, using angled conveyors to transport the pieces to the elevated floors. Mixing may be on a floor above the front end of the ovens or on grade, using elevators to raise dough troughs for dumping into the forming stages. Because the ovens may be used for more than one product, it is common to have roll in roll out units for forming.

3.3 Examples

1. A new snack product was based on using the same dough as an older existing product but with a unique and distinctive shape, achieved by extrusion through an intricate die. The dies were relatively expensive and had a tendency to plug, leading to defectively shaped pieces. A major element of introducing the new product to several plants was training operators in product evaluation. The product did not last long in the market, probably because it provided no distinct consumer

benefit, though it was clever and sophisticated to manufacture. The consumer neither knew nor appreciated what went into it. Discuss what makes a product successful and what leads to failure.

2. Formed, baked snacks have the advantage that they can be stacked like coins and packaged much more densely than randomly dumped chips in a bag. However, they must be cooled first so that they do not create a vacuum in the container, as they would if packed hot. One approach is to stack them while cooling and to season by dropping powdered seasoning on the edges of the stacks. This approach to seasoning leads to non-uniform adherence. Electrically charging conductive seasonings (conductive because they contain salt) helps, but not all seasonings have salt in them. An approach that has helped in some other cases uses an oil mist to improve adherence. Discuss what amount of oil might be appropriate to cause adherence to both rough and smooth snacks.

3. Higher moisture snack cakes, such as muffins, do not stale in the usual manner because they are soft to start with, but may fail because of mold growth. One approach that has worked is the use of controlled or modified atmosphere packaging (CAP or MAP). CAP or MAP reduces the oxygen in the headspace of a package by displacing it with carbon dioxide. Determining the correct atmospheric composition is a matter of trial and error. Once the composition is established, it is applied on special packaging equipment. The packaging material must have moisture, oxygen, and carbon dioxide barrier properties, often achieved by adding an aluminum foil layer between polymer film layers.

4. Salty snacks are often transported from frying and seasoning to packaging by vibratory conveyors, which are used because they can be easily cleaned and are gentle with the fragile products. They also can provide a flexible distribution system by opening and closing gates. It is desirable to minimize residence time while products are exposed to air because they can pick up moisture and begin oxidation of the residual oil, leading to rancidity.

Packaging of salty snacks can range from 1 oz to 1 lb. Oversimplifying somewhat, packaging machines operate at about the same number of packages per minute, regardless of size. This is usually about 100 packages/min. As an exercise, calculate the number of machines needed to absorb the output of one 5,000 lb/h frying line, assuming various distributions of package size demands. The extremes are all 1 oz or all 1 lb packages, neither of which case occurs in reality. Consider how a network of vibratory conveyors might be arranged to provide access to the correct packaging machines for various cases. Now consider how a packaging department might be organized in a multi-line plant, given that packaging machines are relatively independent of specific products. (They must be cleaned between runs of different flavors, to prevent cross-contamination.)

This problem, which is common to many situations where products have several package sizes, is an opportunity to take advantage of diversity, which holds that parallel production lines rarely are synchronized in their demand for utilities or services such as packaging.

5. Estimate the refrigeration load for a bread cooler processing twelve thousand 1-lb loaves per hour. Air is cooled to 75°F. The specific heat of bread is 0.7 BTU/(lb F). Air warms by 20°F. The bread is cooled in 75 min. What is the air flow in SCFM?

6. Thinking about the rounder discussed earlier, for 12,000 loaves/h, what are the approximate dimensions of the dough ball for a 1-lb loaf? (Remember that the dough roughly doubled in volume during proofing.) Assuming a proof time of 8 h and that density continues to change at the same average rate, how much will the volume of a dough ball for a 1-lb loaf change during the time a 2,000-lb trough is running through the rounder. How much will the dimensions change?

3.4 Lessons

1. Water is often wasted in food processing and often can be conserved by re-use, as in the case of steeping and washing corn-based snacks.

2. Packaging capacity following continuous manufacturing processes must be over-sized to avoid costly interruptions.

3. Diversity of demand occurs in most processes and can be used to reduce capital investment.

4. When internal rate processes dominate, as in baking heat transfer, increasing external rates has little effect.

5. No matter how clever the product design and sophisticated the process, only if a product delivers a benefit to the consumer will it succeed.

6. Efforts to automate and mechanize food processes must carefully negotiate the balance between human craftsmanship and human error.

7. Batch processes can be more flexible than continuous processes.

8. Some flavors just take time to develop.

Chapter 4
Breakfast Cereals

Ready-to-eat (RTE) breakfast cereals were among the first deliberately conceived "health foods" developed at sanitariums in Battle Creek, MI, to make more palatable the diets based on whole grains that were advocated then. Some of the processes and products developed in the early days of the industry are still in use. Early products included flakes made from corn grits and wheat berries, shredded-wheat biscuits, and nuggets of barley and other ingredients baked in loaves and then milled and dried. Today, RTE cereals are profitable and diverse segments of major food companies, including General Mills, Kellogg, Kraft (Post, recently sold to Ralcorp), and PepsiCo (Quaker).

A fifth player in the market collectively is the category of private-label cereals made by several small- to medium-sized firms. The private-label products are usually imitations of the branded products but sell for less because the companies have lower marketing costs, do not invest as much in research, and sometimes use less-expensive processes.

4.1 Processes

The major cereal processes are flaking, gun puffing, shredding, and extrusion. As described in Fast and Caldwell (2000), the early cereals had separate mixing, cooking, forming, drying, and coating steps. Typically, many of these process steps were, and often still are, batch operations. For example, whole wheat berries or corn grits are cooked in pressure vessels with added malt, salt, and other ingredients, using direct steam injection. The objective is to gelatinize the starch and temper the grains with enough moisture to make them plastic without turning them to soft mush.

Flaking. After cooking, the grain is cooled and slightly dried before forming by passing between polished steel rolls to form flakes. Whole-grain rice is also cooked and then lightly "bumped" between rolls before forming by toasting in an oven to make a crisp rice cereal. Modern extrusion technology has allowed manufacture of a credible imitation by direct extrusion (Mercier et al. 1989).

In the traditional flake process, flakes from the rolls are dried and toasted in rotary or conveyor ovens. Some modern ovens use high-velocity jets to fluidize the bed of flakes and shorten the drying time. If flakes are to be coated with sugar or other

J.P. Clark, *Case Studies in Food Engineering*, Food Engineering Series,
DOI 10.1007/978-1-4419-0420-1_4, © Springer Science+Business Media, LLC 2009

flavors, they pass through a coating drum or belt and then are dried again. Coated or uncoated flakes often have a vitamin emulsion sprayed on them as a final process step, followed by another drying step, and then packaging. The vitamins and minerals, if used, are added last because they would not survive the high-temperature treatments of earlier process steps.

Cereals are seen as desirable vehicles for delivery of added nutrients because they can represent a significant portion of a daily diet, are usually eaten with milk, and have wide appeal. Often, nutrient fortification is an important part of a given brand's image and marketing message.

Gun puffing. Gun puffing starts with cereal flour or whole grains, depending on the products. Quaker and others make puffed whole wheat and rice, while General Mills uses oat flour and other ingredients to make Cheerios. Flour-based products are formed using a low-pressure extruder similar to those used for pasta or "half-product" snacks. (Half-product snacks are low-density puffed pieces made by frying pellets of starch and flavors.) The pellets are dried and then heated in a closed pressure vessel. The pellets are released from the elevated pressure either by quickly opening a closure or by transfer through a special valve. The sudden pressure change in the softened pellet causes expansion of water vapor and quick cooling, so that the pellet expands and then hardens. The porous shape retains a crunchy texture even in milk. Puffed pieces are dried further, coated and enriched as mentioned above, and then packaged.

Shredding. Shredding involves cooking whole grains to a precise moisture content and then passing them through batteries of special rolls in which one roll is grooved and presses against a smooth roll. High pressures are developed as the grain is forced into the grooves, so the rolls tend to be short, to control deflection. In a commercial line, there are many banks of rolls. The threads of wheat are laid over one another in mats and cut into biscuits. A variation is to have cross grooves so that threads are attached to one another.

In some processes, cooked grains may be stored in a relatively moist state before forming. It may be that this step contributes to final flavor by permitting chemical and biochemical reactions to occur. In general, the traditional, relatively slow batch processes seem to produce the more-complex flavored products, though there are constant efforts to increase speed and efficiency (Valentas et al. 1991).

At least one RTE cereal is made by baking a very heavy loaf of dense bread, chopping the loaf into pieces, drying the pieces, then grinding them further, and packaging. The resulting product is so stable that it can be packaged in a cardboard box with no liner.

Extrusion. One of the great improvements in cereal processing was the application of cooking extrusion. An extruder can combine in one machine the operations of mixing, cooking, forming, and expansion. Quaker was able to focus on extrusion, in part because it had little investment in more-traditional processes other than gun puffing. At the time that Quaker expanded its cereal business, there were no extruders intended for foods, so Quaker used equipment from the plastics industry. Today, there are multiple manufacturers of food extruders.

Since extruded cereals undergo significantly shorter processing conditions (high temperature, short time), they tend to develop different flavor profiles compared to traditionally cooked and processed cereals. Efforts to match the flavor of one with the other are known to present technical challenges. One of the goals of the branded cereal makers is to deliver products with good taste while also delivering healthy nutrition. Lately, this has meant a focus on whole grains as well as lowering the sugar content and increasing the fiber content of cereal products without compromising taste.

A number of companies provide extruders and other equipment to the cereal industry. Some of the challenges regarding extrusion of RTE cereals include requirements for increased sanitation of equipment, driven by concerns about allergens and by the focus on nutritionally functional ingredients. Increasing fiber content in new products, for example, results in larger and more fragile flakes. Flakes can be made by extruding pellets and then flaking them. This permits use of flour and incorporation of other ingredients, as opposed to using whole berries or grains.

4.2 Coating and Inclusions

Sugar coating is a complex topic in its own way. Coatings may be transparent or opaque and may now include artificial sweeteners such as aspartame and non-cariogenic materials such as sugar alcohols. High solids coating solutions must be kept hot and are subject to discoloration. Application can be challenging because the solutions can crystallize and harden quickly.

Fruits, such as raisins, have been added to cereals for many years, but in the search for new products, other inclusions such as nuts, freeze-dried fruit, yogurt, and granola (a cereal mix itself) have been added. These create processing challenges in feeding at correct rates, prevention of segregation, matching moisture activity, and preventing creation of fines.

Raisins and cereal flakes, for example, are quite different in water activity, so that when enclosed in a sealed package, the flakes can become soggy and the raisins dried out. Early packaging used waxed paper in cardboard boxes, which actually permitted some moisture to escape. When cereal packaging was changed to polymer liners, which are better moisture barriers, the equilibration issue became more noticeable. One solution has been to add solids to the raisins, such as sugar or cereal fines coatings, to reduce water activity.

Fines are an almost inevitable by-product of cereal manufacture and have become a useful ingredient in other products, such as breadings and coatings.

Ready-to-eat cereal is an interesting food category not only for the variety and challenges of its many processes, but also for the several ways in which it can be categorized. There are cereals for adults and for children; sweetened and unsweetened; short-lived novelties and 100-year-old standards; originals and copies; whole-grain and flour-based; branded and private-label. For manufacturers, RTE cereal has been profitable, challenging, and a source of technical accomplishment. For consumers, it continues to be nutritious, enjoyable, and generally a good value.

4.3 Examples

1. In the early 1980s, a significant issue in considering new cereal manufacturing facilities was the degree to which the process should be controlled by computers. Computers for process control were a relatively new idea at the time – today they are taken for granted. The owners of one very large expansion determined not only that their new process would be highly automated and controlled by computer but also that there would be no manual back up control capability. This was viewed as courageous, maybe even foolish at the time. An example of the issues that better control was to address is the setting of flaking rolls clearance. Skilled operators adjusted the settings of flaking rolls by listening to them. Each operator had his or her favorite procedures, often quite different from those preferred by others. It was observed that at almost every shift change, there would be a temporary process disruption while the new operators adjusted settings that had been working just fine. Eventually, the process settled to a new condition, until the next shift change. In the meantime, some off-specification product would be made. It was hoped that computer control would reduce such upsets by preserving and maintaining optimal process settings.

   The plant in question took several years to reach satisfactory operation, in part because the extent of training required was much greater than anticipated. Some other forward-looking features were not successful and were replaced. Eventually, however, the plant became the most efficient and lowest cost producer in the entire firm.

   Consider what are the control points in a cereal process? What information would you want to collect? What sensors would do that for you? What would you do with the information?
2. The issue of continuous as opposed to batch processing arises in many food processing situations. Continuous processing has the advantages of maintaining constant conditions and of being relatively stable. However, continuous processes are often less flexible than batch processes. In at least one case of a large breakfast cereal plant, there was a debate between an expensive continuous cooking unit and a battery of conventional batch cookers. Only a limited number of products could be made with the continuous cooker, but it made these very well. The continuous cooker was installed but subsequent market demand favored products that had to be cooked in batches. Consider the benefits and deficits of batch and continuous processes. (Subsequently, a member of the owner's team admitted that they had made an incorrect decision, against the advice of their engineering consultant. It is extraordinarily rare for a consultant to ever hear such an admission.)
3. Fruit, such as raisins, are frequent additions to breakfast cereal products. As previously mentioned, there is a mismatch between the water activity of dried fruit and that of cereal flakes. Determine the respective water activities of the components. Determine the respective weight fractions. (Look at labels.) What will be the equilibrium water activity of the mixture after enough time? (See Rockland and Beuchat 1987, Iglesias and Chirife 1982, among others.) What is the impact

on quality of using freeze dried fruit? What is the impact on cost? (Look at prices in the market.)

4. Traditionally, breakfast cereal plants were constructed in multi-floor towers to take advantage of gravity flow. As processes became more automated, there would be relatively few people working on each floor. Another layout option is to have all the equipment at or near grade, but this then requires several elevations of materials, some of which are difficult to handle, such as cooked grains. An intermediate layout concept is a space frame or sturdy structure within a large room. Communication among operators is improved in such an arrangement, but special care must be taken in the design, especially for flaking or bumping rolls, which can generate significant vibrations. Some handling issues have been addressed with innovative use of pneumatic conveying, in which some conditioning of cooked grains occurs by use of tempered conveying air. Consider the various processes described earlier. What are some of the material handling issues that may occur, and where in the process do they occur?

4.4 Lessons

1. Some food categories have many alternative processes, about which it is difficult to generalize.
2. As previously recognized, some flavors just need time to develop, so high-temperature, short-time heat processes may not give the same results as a long-time, lower temperature process. This may be good or bad, depending on the product.
3. Computer control and automation, now taken for granted, require more highly trained operators than more manually operated processes do.
4. Water activity is an often overlooked parameter that affects product quality and safety subtly.
5. Material handling of solids, especially of those that are wet or fragile, can become a dominant issue in process development and layout. Gravity flow is often preferred.

Chapter 5
Pet Foods

Pet foods have often seen the initial use of technologies that later found application in human food, such as meat analogs, intermediate moisture foods, extrusion, and the use of novel ingredients. Pet foods are often the only source of nutrition for animals and so they require a thorough understanding of nutritive requirements and precise delivery of each dietary component. Pet food processing has evolved from an animal feed mill approach to a human food approach, in part because it has been realized that deliberately or inadvertently, pet foods are consumed by humans.

5.1 Complete and Balanced Nutrition

Companion animals and zoo animals depend almost exclusively on the feed they are given by humans. This means that pet foods are tailored to the nutritional requirements of the species for which they are intended. Cats and other felines are carnivores (meat eaters), so their foods are based on animal protein, while dogs are omnivores (meat and vegetable eaters), so their foods can be cereal based, though they like meat. Vitamins, minerals, and other nutrients are added in sufficient quantities to maintain health in normal animals. Few foods for humans are complete sources of nutrition. (The exceptions are parenteral solutions used in hospitals and supplied intravenously for patients who cannot eat the normal range of foods on which we usually depend.)

While pet foods are formulated specifically for certain classes of animals, it is known that inadvertently or deliberately, they are consumed by humans. A common instance is a toddler trying some of his or her pet's food. Many canned cat foods are essentially red meat from tuna fish and are often bought as an inexpensive substitute for canned tuna. This is one reason that the manufacture of pet foods is properly seen as food manufacture, with all the attendant concerns and requirements for safety and sanitation.

Pet foods are typically regulated by state agencies that confirm the labeling and acceptability of ingredients and formulas. There are ingredients accepted for pet foods that are not approved for human consumption. An example is beef lungs, which are considered so contaminated by dust that they cannot be adequately cleaned for human use. Rendered protein from the manufacture of tallow, grease,

J.P. Clark, *Case Studies in Food Engineering*, Food Engineering Series,
DOI 10.1007/978-1-4419-0420-1_5, © Springer Science+Business Media, LLC 2009

and lard are used in pet foods, but not in human food. Milling and oil seed by-products are additional ingredients for pet foods that are not usually used to feed humans. Ingredients are routinely analyzed for protein, carbohydrate, oil, moisture, and ash before use so that formulas can be adjusted to achieve label declarations. Pet foods are not made by a rigid recipe, but rather are assembled from available ingredients to achieve specified levels of nutrients.

Least-cost formulation. Assembling a formula from various ingredients means that a given product may have varying amounts of each ingredient and, often, different ingredients. If a product is described as having a certain ingredient, then it must have a minimum amount of that ingredient, but if the pet food product has a more generic description, then ingredients can be substituted, so long as the result meets minimum standards for nutrition. The least expensive formula is calculated using linear optimization algorithms, based on current prices and the specific analyses, using nutritional requirements as constraints. An additional constraint on formula is palatability, which is more difficult to quantify.

Palatability testing is conducted using colonies of cats, dogs, and other creatures, less commonly. The dogs are often beagles and the cats are of various breeds. Animals are given their choice of a test formula food and a control and then the amount consumed of each after a fixed time is measured. Multiple animals are used in each test. It was once discovered that great care must be taken in such tests to randomize the order in which tests formulas were presented and to give the animals "days off" between testing. The result is that palatability testing is expensive. The colonies must be maintained and the production of data is slow.

A special category of pet food is prescription diets, which are formulated to address various health issues, such as obesity, dental plaque, urinary tract blockages by mineral deposits, diabetes, and heart disease. Mark Morris was an early researcher in the area of the impact of nutrition on disease conditions. He developed formulas, which initially were intended for home use. Eventually they were made commercially and distributed through veterinarians. While sold in relatively small volumes, compared to maintenance diets, they are highly profitable and must be made very carefully.

Development of prescription diets depends on identifying animals with naturally occurring conditions and then finding formulas that prolong their lives. To do so, pet food companies develop relationships with veterinary clinics at universities and with veterinarians in private practice. They observe their colony animals carefully and treasure those that have a treatable condition. Colony animals normally live their natural life span in the colony, though some may be adopted. They are rarely euthanized. Volunteers or minimally paid people are recruited to play with the cats and walk the dogs. In some pet food companies, there is the charming practice of allowing office animals to roam the halls, sometimes occupying conference room tables or demanding a belly rub.

More seriously, it is important to understand the strict nutritional requirements of pet foods, the variability of ingredients, the potential for human consumption, and the functionality of some formulas to appreciate the design challenges of a pet food process and facility.

Categories of products. Pet foods can be intended for various species – dogs, cats, fish, birds, gerbils, zoo animals, working or military dogs, etc. Dogs and cats are by far the most common and are about equal in volume of food consumed, though there are more pet cats than there are pet dogs. Cats eat less than dogs because they are generally smaller. Pet foods can be dry, wet (canned), or semi-moist. Dry foods are made by extrusion. Canned foods are made by grinding meat and meat by-products, but there are several processes for fabricating slices, chunks, and other pieces from meat and other ingredients. Semi-moist foods are meat and cereal based, using humectants to control water activity while retaining a soft texture (Rockland and Beuchat 1987, p. 369). Humectants are chemicals that absorb water and lower the water activity of a mixture. Sugar, glycerin, sorbitol, and propylene glycol are examples. As previously mentioned, there are prescription diets for a wide range of conditions, but there also are maintenance diets for normal animals; foods for puppies and kittens (who are growing quickly); foods for older animals; and "super premium" foods that make special claims for ingredients, palatability, or performance. An example is dry pet food that has significant amounts of meat, as contrasted with normal dry food, which is cereal based.

Other products for animals include treats, such as biscuits and meat jerky; chew toys made from leather or tanned animal parts (pigs ears, for instance); and the wide range of care products, such as shampoos. The focus here is on the foods, but some treats are made in the same plants.

5.2 Ingredients and How They Are Handled

The major ingredient for dry pet foods is whole corn, which is received in trucks or rail cars of the grain. The grain is unloaded by dumping through grates to a pneumatic pick up, which delivers it to a silo. The silo needs to be large enough to receive a typical shipment while still being partially full. Rail cars can hold about 100,000 lb and trucks hold about 40,000 lb, so silos are sized to hold 150,000 or 60,000 lb, depending on the delivery method. Dry corn is cleaned to remove dirt, weed seeds, and stones and then ground in a hammer mill. Because grinding generates dust, which could explode, and because the mill is noisy, the mill is enclosed in a blast proof room, with blast resistant doors and a blow-out panel to the outside.

The next largest ingredient is soybean meal, a by-product of removing oil from soybeans. The meal is already a fine powder and is normally delivered in bulk trucks and unloaded pneumatically. Other oil seed meals may also be used, such as cottonseed, peanut, canola, sunflower, etc., but soy is most common. Wheat flour by-products are also delivered in bulk. Less heavily used ingredients are delivered in bulk bags, which may hold 2,000 lb or smaller kraft paper bags holding 90 lb. Micro ingredients, such as vitamins and mineral pre-mixes, are delivered in drums.

Meat is usually delivered in cardboard totes holding about 2,000 lb. The meat may be frozen or refrigerated. In either case, it must be stored in a cooler or freezer until used. Some meat by-products may be delivered in bulk tankers and pumped to tanks. These might include liquid fat, blood, and viscera. These materials are perishable and need to be used quickly. Some meat ingredients are delivered in pallets of frozen blocks weighing about 40 lb and separated by slips of paper. It is not uncommon for the blocks to freeze together and for the paper to be stuck to the blocks.

Bulk dry ingredients are formulated by conveying to a scale hopper on load cells. If there is only one scale, only one ingredient at a time can be conveyed. For this reason, it is common to have several scales feeding one collection bin, which in turn feeds a mixer. Corn is conveyed to the mill and the ground corn is collected in a day bin from which it is conveyed to the scale. Bagged and drummed ingredients are often used in integral numbers of bags or drums, which are counted as they are dumped through a bag dump station. Quantities less than a whole drum or bag are scooped and weighed before being conveyed to the mixer. Usually there is a bin beneath the mixer from which the formula is conveyed or dropped to the feed bin of the extruder. Dry mixes are sometimes also combined with meat for wet pet foods. Some mixers have "bomb bay" bottoms so the entire contents can be quickly dropped and the next load started.

Bag dump stations typically have a grate over a small bin connected to a pneumatic transport line. There is a dust collection hood over the grate. Space is provided around the dump for pallets of bags or drums. Empty bags and drum liners may be removed by vacuum conveying or just piled up for later removal. The operator slits the bags or tilts the drum over the grate. The operator should not be required to lift more than 40 lb vertically, so often there is a scissors lift to elevate pallets to the same height as the grate, so that movement is horizontal. Extruders can easily consume hundreds of pounds per minute, most of which is provided by bulk ingredients, but the operator must be able to supply the minor and micro ingredients at a proportionate rate.

Extruders take some time to stabilize and so once they are running, it is best to keep them fed and running smoothly.

Meat ingredients are more difficult to convey. Traditionally, frozen blocks were counted into a grinder/mixer. Pumpable ingredients were pumped to a scale hopper and dropped in. Dry ingredients were pneumatically conveyed or dumped. Once ground, the meat mixture can be pumped to a can or pouch filler.

Fabricated pieces are mixed and ground, but then are cooked in pans to make loaves or as long strips on belts, both going through steam ovens. The loaves and strips are chopped and sliced to make specifically sized pieces, which are then mixed with gravy and filled. Usually pieces are dropped into a container and then liquid is added, since it is hard to keep the pieces evenly suspended. In some formulas, the pieces are largely vegetable protein and what meat is present is in the gravy.

Cans or pouches are sterilized by cooking in batch or continuous retorts, as with other canned foods. The critical element is to achieve a target temperature at the center of the container sufficient to kill spores of *Clostridium botulinum*. (Thermal processing is discussed in greater detail later in this text.)

An improvement to meat formulation is a system in which separately ground ingredients are stored in agitated bins mounted on load cells. The components of a mix are collected in a screw conveyor and transferred to a mixer, from which they are conveyed to the filler. Some issues with this system include possible cross-contamination among ingredients, since a common grinder is used; the difficulty of maintaining frozen meat in a flowable state; the need to properly design the cycles and capacities of the storage bins; and the mechanical complexity of the many screw conveyors, holding bins, mixers, and grinders.

The creation of meat analogs from vegetable protein is more advanced in pet foods than in human foods and is an example of the way in which pet food technology has often preceded technology applications in human food. Other examples include formulation for nutrition and health – a necessity in the case of pet foods; the use of novel ingredients, such as soy meal; and the use of extruders.

5.3   Some Unit Operations

Formulation. The ingredients and how they are handled have been described. One issue is whether to provide an operator with a supply of each ingredient, either in bulk or on pallets of bags, or to have another person prepare a "kit" for a batch so that the person dumping is not also responsible for the recipe. Both approaches are used. Some of the trade-offs include the space required or available near the dumping station, the apparent addition of a job (less than it seems because some one needs to replenish the supply in any case), and the "dumbing down" of the operator's job. The last issue had the greatest effect in at least one case. The suggestion had been to introduce a "kan ban" system, in which an empty pallet indicated that a new kit was needed. The operator merely needed to empty the bags or drums in the kit. Some one else assembled the kit from a large supply of ingredients. The operators preferred being responsible for following the recipe and having a supply of ingredients around their workstation. This required generous floor area. It is important to recognize the human factor – the pride and dignity of the hardworking people running a food line.

Formulating with meat is hard, messy, smelly work. Automation with the grinding, holding, and conveying system previously described not only led to more precise formulation but also relieved people of a job that was unpopular. However, it is a substantial investment compared with a manual system. Canned pet food is declining in volume. Some relatively new canneries have been closed because of low demand. One small boost may come from wider use of retort pouches, which have some consumer appeal because they are easy to open, light in weight, and easier to discard. Recent increases in the cost of corn and soybeans may make canned pet foods more cost competitive with dry pet foods.

Mixing. Dry formulas are typically mixed in large ribbon blenders, holding up to 5,000 lb. Meat mixtures use sturdy mixer/grinders in which the agitator conveys the mix through small holes in a plate past which a rotating knife moves. Meat mixtures are typically quite viscous and difficult to mix uniformly, but mixing too long can raise the temperature from the introduction of energy. Increased temperature can lead to spoilage and to smearing of fat. One consequence is that meat formulas are probably not uniform. (The same issue arises in making sausages for human consumption. Mixtures of meat for human consumption are often left to stand overnight, during which salt equilibrates by diffusion. This is not usually done for pet foods.)

Some dry pet food products consist of several different colored and shaped pieces in the same bag. Each of these is made on its own extruder, dried separately, and then conveyed to the packaging equipment. Larger pieces, called kibbles, do not lend themselves to mixing well together.

Extrusion. Cooking extrusion is used to make breakfast cereals and snacks, but is probably used to a much greater extent in pet foods. It is an efficient way to mix, cook, form, and expand in one operation. Extruders can be single screw or twin screw and may be heated or cooled. Some extruders have preconditioning vessels in which some cooking and moistening occurs. Steam, water, dyes, and other liquid ingredients can be injected and vapors removed through openings in the barrel of an extruder.

A single screw extruder behaves like a centrifugal pump while a twin screw extruder behaves like a positive displacement pump. In a single screw extruder, output pressure declines as flow rate increases. At the same time, the pressure required to flow through a given die plate increases as flow increases. This means there is one operating point for a given material, in a specific extruder, with a specific die plate. The variables for the die plate are the number of holes, the shape of the holes, and the thickness of the plate. There are many variables for the extruder, including the shape of the screw:

- depth (distance between tip of screw flight and root or shaft),
- pitch (distance between flights),
- variation of depth and pitch,
- presence or absence of special sections for kneading or back mixing.

Additional variables include screw rotation speed, feed rate, and temperature profile. The barrel and the screw may be heated or cooled, and this can vary by section. The length of the barrel is another variable, usually expressed as number of diameters (L/D). Barrels and screws are often made as modules that can be added or removed. Heating and cooling may be through jackets or by direct injection of steam or water. At least one model of extruder is heated by electrical resistance.

Heating and cooling affect the degree of cooking that the material experiences. Cooking for starchy materials is related, in part, to gelatinization of the starch from moisture and heat. Proteins, on the other hand, denature under heating. The apparent viscosity of starchy materials tends to decrease with heating while proteins may increase in apparent viscosity with heating (Mercier et al. 1989). Apparent viscosity affects the flow of the dough mass through the extruder and die plate. The decrease in pressure as the dough exits the die causes water to flash to steam and the extruded pieces to expand. This creates a rigid foam, which hardens further in drying. The crunchy texture contributes to palatability. Expansion also lowers bulk density, which means a given weight occupies more volume, leading to larger packages, which appeals to consumers.

A twin screw extruder generates whatever pressure is needed over a range of flow rates. Thus, in contrast to the single screw, it is more flexible and also can feed a wider range of materials. Of course, there is a cost – twin screw extruders are significantly more expensive than single screw extruders for the same capacity.

Drying and cooling. Because pet foods are fairly sturdy particles, they can be piled on a conveyor belt for drying. Most dryers have at least two passes and may have several different zones for heating. Air is heated by burning natural gas and circulated with fans. Most is recirculated while a portion is exhausted by adjusting dampers. After drying, the pieces are cooled, usually by contact with ambient air, either in the last zone of the dryer body or in a separate unit. Using ambient air means that cooling is inconsistent with the seasons – poor in the hot summer and excessive in the winter. Cooling too much may cause condensation of moisture on the pieces, which could encourage mold growth.

Coating may occur before or after cooling. Coating is often with edible fat – tallow or grease. Some firms believe the coating is absorbed better by warm pieces, while others find application to cooled pieces to be adequate. One issue is cleaning of the cooler if fat coating is applied to the warm pieces. Some fat adheres to the cooler belt and is hard to remove. Fat and dry flavors are believed to improve palatability of the foods. Dry flavors include yeast, egg, and some proprietary blends. Some dry flavors have flow improvement aids added because otherwise they tend to cake from absorbing moisture. Flow aids include silicon dioxide and talc. Because the materials used for pet food flavorings are relatively low cost and often made from off-spec ingredients, they can be highly variable in properties. This can have a significant effect on feeders that rely on consistent properties, such as angle of repose. In at least one case, a feeder was designed based on tests on a supposedly representative sample of dried egg. In reality, the purchase specification made no mention of amount of flow aid or of physical properties. Purchased dried egg varied widely from some that would not flow at all to some that flooded and aerated. Needless to say, the purchased feeder could not work and was replaced with a more forgiving design, which, of course, was more expensive.

Packaging. As with some other food technologies, pet food packaging has sometimes pioneered for human foods. Dry pet food was often sold in coated paper bags ranging from 1 to 40 lb. These were typically sewn closed. Coatings were thin and intended to protect against moisture, primarily. Grease and fat coatings of the product often bled through the paper to make an unsightly blotch. One improvement was to use polymer coatings that could also be heat sealed. A still further improvement has been flat-bottomed polymer bags that can display brighter and more vivid graphics while providing better barrier properties.

Wet pet foods were packaged in three piece steel cans of various sizes. Those for dogs were typically tall, while cat food cans were lower in profile. Newer cans are two piece and may be of steel or of aluminum or of steel with easy open aluminum lids. The latest innovation for wet pet foods is retort pouches with gussets that permit them to stand up. (Gussets are a double fold at the bottom that gives the pouch a base as contrasted with "pillow" pouches that have four seals and can not stand on their own. Gusseted pouches use more material and so are more expensive, but have significant marketing advantages. Pillow pouches often require a paperboard carton, which adds cost exceeding the extra cost of a gusset.)

5.4 Examples

1. After design of a new pet food plant had begun, the owner decided that the plant would be organized using a team approach. The plant was to have both wet and dry production areas – one based on extrusion using corn and other dry ingredients and the other using meat and a hydrostatic cooker for sterilization of cans. The essential idea of a team approach is that there are fewer layers of management and supervision, that people are cross-trained, and that the teams or small groups organize themselves and solve their own problems with little oversight. What would be the architectural consequences of such a conceptual decision? One belief was that the teams should have a meeting and work space that was close to where they normally were stationed. Another belief was that there should be few barriers to communication among teams and team members. A solution that evolved was to construct a spine through the building that naturally divided the two types of manufacturing space, had windows out on to the manufacturing floor, and provided open work spaces with computers, white boards, and conference tables for each team. These spaces were distinct from those assigned to the plant management and to common services such as the cafeteria and fitness center (itself a relatively novel concept for a plant). Discuss other influences on the architecture of a food plant. How might this pet food plant be designed today? (The real plant in this case was runner up for Food Plant of the Year in spite of not being a human food plant.)

2. Consider the meat formulation system described. This has ten different meat ingredients, each of which arrives as frozen 40-lb blocks on 2,000 lb pallets. Total formula use is 18,000 lb/h. The worst case uses six of the ingredients at 15% each in the formula. (The balance is cereal and water.) At what rate must the primary grinder operate to keep the holding bins adequately full? Assume the conveying system can run at 50 lb/min. Assume the grinder does not need to be cleaned between ingredients and that the grind plate need not be changed. (These assumptions are not always true.) How big should the bins be? What is the consequence on grinder size and rate if it needs a 5-min water wash between ingredients? What if it needs a longer cleaning? How long a cleaning impacts the design – that is, if not 5 min, how many minutes before something significantly different must be done, such as getting a second grinder? (Assume the biggest grinders can do 20,000 lb/h of most meats.)

3. A pet food company was persuaded to order two twin screw extruders as "the newest technologies." Their engineering consultants were unconvinced. Eventually, the company canceled the order, despite having to pay most of the cost of the first unit and getting nothing for it. Replacement single screw extruders cost 40% of the cost of the twin screws. Was this a good decision? What do you need to know to reach an opinion? Assume that single screw extruders cost about

$1 million. They were the company's current technology, that is, they made all the current dry products. The project was under severe budget pressure, so even a small savings in capital was helpful.

4. This is not a technical lesson, but it occurred in the context of pet food projects. An engineering firm was executing several major projects for the same pet food client at the same time. One was proceeding relatively smoothly, but the other had encountered numerous problems, leading to cost overruns and schedule delays. The problems were complex and, while not entirely the fault of the engineering firm, it certainly shared some blame. The engineering firm operated on a cash basis accounting and the end of its fiscal year was approaching. The client had not paid several hundred thousand dollars of invoices. This income was critical to the engineering firm booking a profit for the year. The delay in payment had occurred for several months with no explanation. A senior executive of the firm (you might correctly guess it was the author) was dispatched with a more senior colleague to discuss the problem with the client. The more senior colleague missed the plane, leaving yours truly to face an unhappy and behind-in-his-accounts client alone.

The client's position was that both parties shared some blame for the problems and that the engineering firm's share came to just what was owed and being withheld. The engineering firm was rather desperate to get some already anticipated cash before closing its books – what happened later was of less concern. (One has to understand how a small company's relationship with its banks depends on meeting projections for cash flow, not unlike the expectations of the stock market for publicly listed firms.)

It was eventually suggested that the client pay the amount in dispute immediately with the understanding, in writing, that the engineering firm would apply a discount to future work until the disputed amount had been repaid more than in full. One consequence, the reader will immediately see, is that it was in the client's interest to continue employing the engineering firm so as to get the full benefit of its concession. The engineering firm paid for its role in the troubled project without really conceding any fault, but realizing a reduced profit on some subsequent work. The relationship continued for some years because of a feeling that both sides had been fair to each other.

5. I have been in more than 200 food plants in the course of my career. A fair number I had a hand in designing, in others I have worked and in others I have toured as a visitor, sometimes professionally and sometimes as a tourist. One of the best and one of the worst food plants I have ever seen were both pet food plants. The best was run by a person who had been manager of a breakfast cereal plant and quickly saw the similarities between breakfast cereal and pet food manufacture. Both segments used cereal ingredients, both used extrusion and other unit operations, and both faced similar challenges of dust, infestations, and material handling. This manager was scrupulous about sweeping up spills, keeping the plant floor uncluttered, and keeping equipment maintained. He considered his plant a food plant. Some years later, I won a major assignment for my firm with another pet food company by promising to design a new facility as a food plant.

Their previous dry pet food plants had been designed more like feed mills, with slip form concrete silos, steel truss roof supports, and carbon steel equipment.

In contrast, the other pet food plant to which I referred had pallets of thawing meat ingredients stacked on the floor, dripping vile liquids. They had no coolers or freezers. Much of the equipment was made of carbon steel and showed signs of corrosion. The space was poorly illuminated. The management's attitude was that since the product was canned, they need not worry about contamination because everything bad would be killed by the severe thermal process. Some human food canneries still hold that same opinion. Anyone who cared about his or her pet would be quite unhappy after seeing the second plant.

Consider the philosophical and business implications of the design and operating approaches represented by these two extremes. There are those who say things like "They are only animals," "We are using by-products," and "We need to be the low cost producer." Others say, "A kid in Puerto Rico might eat this stuff we make," "Some companion animals are almost like children to those who care for them," and "We can make a profit by selling quality." Where do you stand? There are no right or wrong answers, though you can probably guess my opinion.

5.5 Lessons

1. Pet foods have pioneered many food technologies, including nutritional formulation, meat analogs, extrusion, and the use of innovative ingredients. It is a fruitful area in which to seek new ideas.
2. Pet food plants should be designed and operated as if they were human food plants, even if regulations do not literally require this.
3. Creativity can be applied to business practices as well as to technical problems.
4. Win–win, equitable solutions to relationship problems are best in the long run.
5. Most business relationships will encounter problems at some point. How the problems are resolved will dictate whether the relationship continues or not.

Chapter 6
Fruit and Vegetable Juice Processing

6.1 Citrus Fruit Juice

Single-strength orange juice that has not been concentrated (not-from-concentrate, NFC) is the most rapidly growing consumer product made from citrus fruit. Historically, frozen concentrated orange juice (FCOJ) was important, and concentrate for industrial uses is still a major product. Single-strength juice made from concentrate is the most popular consumer use. Much of this is made in local dairies for efficient distribution under refrigeration.

Florida and Brazil are the two most important sources of oranges. Citrus crops are also grown in Texas and California, as well as Israel, Spain, Australia, and South Africa. About 95% of the orange crop in Florida is processed, with the balance sold fresh.

6.1.1 Processing Steps

Oranges are picked from their trees by hand (although increasing amounts are harvested mechanically) and delivered to the plant in bulk trucks. There are standards for fruit grades, and prices are determined in part by yield and solids content. A sample is taken from each load, and the juice is extracted to determine yield and juice properties. Soluble solids, mostly sugar, and sugar-to-acid ratio are important parameters, which can vary with the time of year and variety. Color is also important and also varies. Juices from early and late in the season are often blended to standardize color and flavor. Yield of juice is generally about 50%, so profitable utilization of the remaining portion is critical.

Oranges are stored in bulk bins until needed. The outside surface is washed with a detergent and warm water, and then the oranges are sorted by size. The machines used to extract the juice are size specific. There are three approaches to extracting the juice.

One approach utilizes a reamer using the same principle as a kitchen juicer, namely, cutting the fruit in half and pressing each half against a rotating burr or reamer. On the commercial scale, each machine has pockets of a specific size

J.P. Clark, *Case Studies in Food Engineering*, Food Engineering Series,
DOI 10.1007/978-1-4419-0420-1_6, © Springer Science+Business Media, LLC 2009

arranged in a circle. Each fruit is cut in half and then pressed against a reamer. The peels are collected for further processing as the juice flows into tanks.

Another approach is the squeezing extractor in which the fruit is penetrated by a porous tube and then crushed by interlocking fingers. The juice flows through the tube, and the shredded peel is collected.

Both machines are in wide use. The major difference between them is the form in which the peel is recovered. Issues that can affect extraction include yield and the amount of peel oil that is in the juice. There are limits on the peel oil content of juice because the oil is bitter. Some oil is necessary for good flavor, but too much is a defect. Also, the oil is a valuable byproduct in its own right (Tetra Pak 1998).

The third approach to extraction is usually applied to other fruits where peel oil is a primary product, such as lemons in Italy. In this approach, the peel is removed from whole fruit by abrasion, and then the fruit is crushed for its juice. The resulting juice is inferior to that produced by the reamer or squeezing machines and may be less sanitary.

Juice is normally passed through a finisher to remove seeds and membrane, or rag. This operation can also control the amount of pulp in the juice. Commercial juices are offered with various levels of pulp or juice sacs. The pulp removed from low-pulp products can be added to those with higher levels.

Juice is pasteurized to reduce pathogenic and spoilage microorganisms and to eliminate enzymes that can inactivate pectin and thus cause separation of pulp from serum (the clear fluid portion of juice). Pectin is a naturally occurring hydrocolloid that contributes to the desirable mouth feel of juice and helps keep pulp in suspension.

Single-strength pasteurized NFC juice is stored aseptically in bulk under refrigeration for later packaging. Juice may be shipped under aseptic conditions in trucks, rail cars, and ships carrying millions of gallons. Consumer packages include single-serve aseptic boxes, gable top coated paperboard cartons, glass bottles, and plastic jugs. Most are sold refrigerated, but aseptic packages are shelf stable.

The thermal treatment for NFC juice is less severe than that for juice that will be concentrated because there is additional heating in the evaporation process. This is one reason NFC is considered to be a superior product as compared with reconstituted juice made from concentrate. The usual evaporator for making concentrate is the "thermally accelerated short time evaporator" (TASTE), which has seven effects and eight stages. The heat is exchanged in long vertical tubes. Temperatures start at about 100°C and are reduced while water is removed until the soluble solids are about 65%. (Soluble solids are normally expressed in degrees Brix, and measured by refractive index; 1° Brix is 1% solids.)

Some competing evaporators use plate heat exchangers, and are also multi effect, meaning that the steam evolved in one stage is used as the heat source in another. This is achieved by lowering the pressure in each successive stage. The result is that one pound of steam, applied in the highest-temperature stage, can remove several pounds of water from the juice, reducing energy costs. A mechanical vapor recompression (MVR) evaporator compresses the evolved vapor to serve as the heat source. This achieves the equivalent of many effects in one stage and uses electric power instead of steam from a boiler.

The high-solids concentrate may be sold as is, around 65% solids, or it may be diluted to about 42% with water and a commercial flavor mixture (derived from essence and peel oil) for consumer packaging as the standard 3:1 FCOJ. Essence is concentrated flavor removed in the first stages of evaporation, then concentrated separately. The chemicals contributing to orange flavor are a complex mixture of higher alcohols and esters, which normally would boil at temperatures higher than the boiling point of water. However, their solution with water is highly non-ideal thermodynamically, so they can be stripped from water by distillation. Most essence is sold to the flavor industry, where it is manufactured into flavors for use in commercial retail juices and beverages.

6.1.2 By-products

As is true of many other foods, by-products are critical to the economics of citrus processing. Besides essence, some of the significant by-products are dried citrus pulp pellets (used for cattle feed), molasses (used in animal feed and beverage alcohol fermentation), pectin, peel oil (used in flavors), and d-limonene.

The major by-product of citrus processing is cattle feed and molasses made from the peel. The peel is first pressed, and the press cake is dried and pelletized. The press liquid is concentrated to make molasses, which may be added back to the dried peel or sold separately as a feedstock to make beverage alcohol by fermentation.

Cold-pressed peel oil is recovered by extracting the oil from the peel by contacting with water, resulting in an emulsion. The oil is separated by centrifuging and is sold, mostly to flavor companies, who fractionate it into various components, most of which find their way back to the juice companies.

During the evaporation process to concentrate the molasses from the press liquor, oil that remained in the peel is recovered. This oil is composed of terpenes, largely d-limonene, which is recovered by distillation and sold as a solvent or as a feedstock for chemical synthesis of terpene resins.

Pulp from the finishers may be washed with water to recover extra solids. Usually there are at least three stages of countercurrent washing, in which fresh water first contacts the exhausted pulp and then proceeds stage by stage until it exits after contacting the fresh pulp. The pulp and wash are separated by finishers. The wash water is concentrated to make water-extracted soluble orange solids (WESOS). This material cannot be added to pure juice but is used in other products, such as soft drinks (Fig. 6.1).

There are many other possible by-products of citrus. For example, there has been interest in the flavonoids of citrus, mostly minor organic compounds, which have various activities. Each citrus fruit has characteristic flavonoids, which contribute to its distinct flavor in some cases. For example, tasteless hesperidin is found in sweet oranges, bitter naringin is in grapefruit, and neohesperidin is in bitter oranges. Some

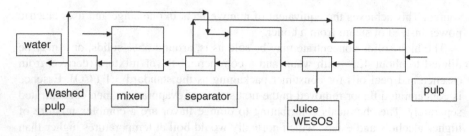

Fig. 6.1 Counter-current extraction pulp washing system

of the dihydrochalcone derivatives of these compounds, made by hydrogenating them, are intensely sweet. The sweetness, however, also lingers, so their applications in foods have been limited. However, at least one, neohesperidin dihydrochalcone, is approved for food use and works well in chewing gum, where the lingering flavor is an advantage.

There is recent research to find biological activities for the minor constituents of citrus fruit, but commercial application may be limited because of the expense of extraction and purification of minor components from large amounts of peel residue, requiring treatment or disposal of large waste streams. The pharmaceutical approach of organic chemical synthesis would probably be more efficient for any component with true biological activity.

6.2 Other Juices

Other popular juices include apple, grape, tomato, cranberry, pineapple, blends of these, and juice-based beverages that may have 10–30% juice equivalent. In general, juice is extracted by crushing the fruit or vegetable and then pressing the pulp through filters or screens. Crushing or shredding may be performed in a hammer mill or other similar device. Pressing can be in a screw press, a batch hydraulic press that uses filter cloths wrapped around pulp and stacked between plates that can be moved to compress the pulp, and other devices that use perforated metal screens and may use air pressure to compress the pulp. Where the juice is concentrated, the pulp is often washed with water to extract additional solids. Hydrolytic enzymes are sometimes used to help release additional juice from plant cells.

Enzymatic activity can be helpful or harmful and so needs to be understood. Naturally occurring pectin methyl esterase breaks down pectin when cells are ruptured. This can reduce viscosity in tomato products such as catsup and sauce and can increase sedimentation of solids in pulpy juices. Other enzymes, as mentioned, can help release juice from cells. Most juices are pasteurized by heating to inactivate enzymes and to kill pathogens. Because most juices have a pH below 4.6, they can be heated to about 180°F and held for about 30 s. Tomato juice requires a more rigorous treatment to reduce flat spoilage microorganisms.

Vegetable juices often have higher pH, but can be preserved by adding acid, usually as vinegar. This is a process called pickling. The products are used as ingredients in other foods, such as salad dressings, sauces, and blended beverages. Juices may be cloudy or clear, depending on whether they are filtered to remove suspended particles.

6.3 Bulk Aseptic Storage

Bulk aseptic storage refers to processing juice during a harvest season for later packaging or other use. The process involves heating, holding, and cooling juice or puree and then filling into a sterile container under sterile conditions. The containers may be single-serve coated paper cartons, plastic bottles, polymer bags 5–300 gal, or metal tanks. Each of these is sterilized in a different way. Coated paperboard is contacted with hydrogen peroxide and then heated to decompose the hydrogen peroxide. Plastic containers may be contacted with hydrogen peroxide or sterilized by the heat used to form them by blow molding. Bags are sterilized by exposure to ionizing radiation. Large storage vessels are sterilized by filling with a liquid solution containing iodophores or other chemicals. The solution is displaced with sterile nitrogen, which is sterilized by passing through very fine filters.

The filling environment is sterilized with steam and then maintained by pressurizing with filtered air. Some filling areas use laminar airflow to maintain sterility. Laminar airflow means that air is introduced gently at the top of the space and that there are no obstacles that could cause turbulence. The airflow removes any potential contaminants. Aseptic connections for filling and emptying are sterilized by washing with alcohol or by heating with steam.

Bulk aseptic storage prevents spoilage at room temperature, although sometimes cooling is applied. There are ocean going ships with multiple 1 million gallon aseptic tanks used to transport orange juice from Brazil to Europe and the United States. The juice is repasteurized upon unloading and before filling into consumer packages.

6.4 Examples

1. During ocean transport of bulk juice stored aseptically and refrigerated, the refrigeration system failed. Not realizing that the juice would have been fine without refrigeration, the crew opened the tank and added ice, thereby contaminating the juice. This incident led to the imposition of juice HACCP guidelines, which require that juices be given a pasteurization treatment before filling, even though, in theory, aseptic juice could be unloaded under sterile conditions and safely packaged. Other instances in which "fresh" fruit or vegetable juice was responsible for cases of food borne illness have led to the requirement that all juice be given a treatment that results in at least a fivefold reduction in pathogens

of concern. This is normally a thermal treatment, but can be achieved in other ways, which are discussed in greater detail later in this book. Oranges specifically can be treated by careful hand picking from the tree and surface washing with warm water and detergent to make fresh juice without thermal treatment. Most other juices require thermal treatment.

2. HACCP, mentioned in the previous example, stands for Hazard Analysis Critical Control Point. It is a seven-step procedure for identifying in advance potential biological, physical, and chemical hazards to health from food manufacturing and applying methods to prevent the hazards. Those methods have critical control points, such as temperatures and times, which must achieve certain values, based on scientific evidence, to achieve their preventive purpose. The HACCP plan identifies the hazards, identifies the critical control points, and documents monitoring procedures and means of responding to deviations from the critical control points target levels.

    Consider what might be critical control points in juice manufacture. List as many as you can. Discuss which are truly critical. Why might one want to limit how many critical control points are identified formally? In addition to a mandatory HACCP plan for juices, a manufacturing plant also needs Sanitation Standard Operating Procedures (SSOP) and must otherwise adhere to Good Manufacturing Practices (GMP). Consider how these differ from one another.

3. A dairy that co-packed juices made from concentrate in single-serve aseptic cartons experienced some spoilage in certain products. Several explanations were considered. First, the water used for reconstitution was highly contaminated. While the juice was thermally processed, the objective was to achieve a certain level of reduction from the original level – say fivefold. If the original level of microorganisms was high, then after reduction, it still could be high. The plant operators dismissed this possibility. Nonetheless, it would have been better practice to treat the water by filtration before use. A second possibility was that residence time in the hold tube was different from that expected because the various juices had different viscosity from that used in design. Orange juice was used in design, but the spoilage occurred in pineapple juice. In particular, if turbulent flow were assumed in design, but laminar flow was experienced in practice, the hold tube would be too short and thus the treatment inadequate. Consider what changes in flow rate or viscosity might cause flow conditions to change from turbulent to laminar. (Reynolds Number is the relevant parameter.)

$$N_{Re} = \rho V D / \mu \tag{6.1}$$

where

ρ is density, kg/m^3
V is velocity, m/s
D is pipe diameter, m
μ is viscosity, Pa s, kg/m s

Reynolds number below 2,100 indicates laminar flow; a value above 4,000 indicates turbulent flow (Valentas et al. 1997, p. 8). The region between these two limits is called the transition region, for which it is difficult to calculate the velocity profile. Good practice assumes laminar flow in this region for the purpose of estimating residence time distribution. (This topic will be revisited in a later chapter.)

A third possible explanation came to light later. The same heat exchanger and hold tube were used for processing and packaging other beverages, including some that were dairy based. Between runs of different flavors, the system was cleaned-in-place (CIP) using an alkaline solution. The system was left full of rinse water, which was heated up to sterilize the equipment before the start of the next run, and then discarded. The first volumes of juice were slightly neutralized by the residual basic solution and thus made more vulnerable to spoilage because their pH was increased. Good practice in CIP is to rinse with water whose pH is adjusted to resemble that of the next product scheduled to be run. This requires addition of acid or base and measurement of pH, neither of which are common on most CIP systems.

4. A pulp washing system in an orange juice plant had two stages of mix tanks and finishers. What is the benefit of adding a third stage? This is a good opportunity to apply a very useful tool called the Kremser equation.

$$N = \log\left[(y_b - y_b^*)/(y_a - y_a^*)\right]/\log\left[(y_b - y_a)/(y_b^* - y_a^*)\right] \quad (6.2)$$

where

N is the number of stages, dimensionless
y is the concentration of soluble solids in the wash liquid
x is the concentration of solubles in the liquid associated with the pulp.
The subscript a refers to the entering solid and the leaving concentrated liquid; subscript b refers to the leaving pulp and entering wash liquid. See Fig. 6.2.
The superscript $*$ refers to the equilibrium value of concentration. In this case, $y_a* = x_a$ and $y_b* = x_b$ (McCabe and Smith 1976).

Some assumptions are necessary:

• The pulp is insoluble.
• The amount of liquid associated the pulp is constant. (This is called constant underflow.)

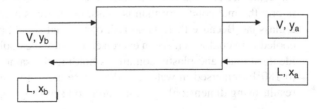

Fig. 6.2 Washing system

- The wash liquid is water with no dissolved solids.
- The liquid entering with pulp is essentially juice with about 12% solids.
- Pulp concentration is about 50%.

The solution relies on calculating material balances around the system (Fig. 6.2). Material balances are valuable in solving many problems in food engineering. What additional information would be useful in this example? For instance, one might like to know the approximate cost of equipment, the value of solubles (check the prices of commodity orange juice in the Wall Street Journal), and the quantities of pulp involved. A typical flow of juice is about 2,500 l/h. Pulp levels are about 25% in the juice from the extractors and this is reduced to about 12% by weight in finishers. There is enough information to estimate flow rate of concentrated pulp to the washer. Another variable is how much wash water to use. Try several values to see the effect on recovery of sugars.

5. A juice manufacturer wanted to consider switching from glass containers to plastic. What are some of the issues? Glass is an inexpensive packaging material with very good barrier properties for moisture and oxygen, but not for light. It also is breakable. On a glass packaging line, it is customary to stop the line whenever there is a broken jar, discard product immediately adjacent – several dozen containers or so, as determined by quality control people, and to put product produced in an hour or so before the break on hold in case there had been other, unnoticed breaks. These production interruptions are costly.

   High-acid juices are hot filled, meaning they are heated to about 190°F, filled into ambient temperature containers, capped, held for about 30 s to allow the hot juice to sterilize the container, inverted to sterilize the cap, and then carefully cooled with tempered water. The temperature of the cooling water is controlled so as not to thermally shock the glass, which could break if contacted with cold water. Glass is also relatively heavy, incurring costs to deliver empty containers and finished products.

   Plastic jars and bottles are made from polyester terephthalate (PET), polypropylene (PP), polyethylene (PE), and some other polymers. PET is inexpensive and very clear. It softens under heating and has adequate barrier properties. Polypropylene is less clear but seems more so when in contact with liquid. Containers can be made with barrier layers within the body and can have various shapes and sizes. Plastic containers are generally lighter than their glass counterparts but may be more expensive because the material is more expensive. Cost and resistance to deformation under heat are directly related to how heavy the container is. Likewise, the time to cool is related to the weight. Plastic is a poorer thermal conductor than is glass, but there is a greater thermal mass with a glass jar. Because there is no risk of thermal shock, plastic containers can be cooled with cold water. As an exercise, measure the cooling curves for otherwise identical glass and plastic containers holding the same amount of water heated to 180°F immersed in water of 100°F. (Be careful with the hot water.) Plot the results using dimensionless temperature on semi-log paper vs. time:

$$(T - T_c)\big/(T_o - T_c) \qquad (6.3)$$

where

> T is average temperature of contents, K (stir with thermometer)
> T_o is initial temperature, K
> T_c is temperature of cooling water, K

Theoretically, the plot should be close to a straight line and the difference in slope should show the effects of the different packaging material.

Another consideration in changing packaging material is the use of existing equipment. Conveyors designed for glass containers rely to some degree on their weight to keep them on the conveyor. Plastic containers, being much lighter, can have a tendency to fly off conveyors. Fortunately, they do not break when they hit the floor. Equipment to apply caps must hold the bottles with a specially molded ring of extra material near the top because the side walls soften with hot filling.

6.5 Lessons

1. Juices are one of the few food categories with mandated requirements for HACCP plans. It is likely that, eventually, all food manufacturers will apply the HACCP approach.
2. Bulk aseptic storage enables the preservation for later use and long-distance transportation of seasonal fruit juice and purees. It is a valuable technology, which has other applications.
3. Fruit processing, especially of citrus, depends on productive utilization of by-products for success.
4. Pasteurization and hot filling enable preservation of good flavor because most fruit juices are high acid. Low-acid vegetable juices are often acidified with vinegar to enable a mild heat treatment.
5. Plastic is replacing glass as a construction material for containers of juices and other foods because of its lighter weight and adequate barrier properties.

Chapter 7
Membrane Processing

Membranes have become important tools in food processing, permitting ultrafiltration, microfiltration, and reverse osmosis with selective polymer membranes. The goal may be concentration, clarification, or fractionation. Membranes are used to clarify liquids, remove impurities from water, concentrate juices, and treat wastes. Flow rates are a function of molecular weight and particle size as well as pore size and thickness of the membrane. Various process arrangements can have dramatic effects on performance.

Membranes differ in their apparent pore size, their affinity for dissolved and suspended matter, their materials of construction, their costs, and their durability. Early reverse osmosis membranes were made of cellulose acetate, valued for its ability to reject dissolved salt and permit water to pass through. Reverse osmosis is now a well-accepted process for desalinating seawater and brackish water for drinking and agriculture.

Other polymers have been applied, which are more durable than cellulose acetate and can have precisely controlled molecular cutoff points, permitting fractionation of suspended and dissolved molecules.

Membranes are available in flat sheets, relatively large tubes (0.5 in. diameter), hollow fibers, and spiral-wound modules. Flow is usually parallel to the membrane surface at relatively high velocity to minimize the buildup of a film of concentrate at the surface.

Membrane equipment is typically fabricated in standard modules, with multiple modules connected in series or parallel to provide sufficient surface area for a given flow rate. The more porous the membrane, the higher the flow through it at a given pressure driving force. Rates can vary from 5 to more than 100 gal/day/ft^2. Of course, at the higher rates, through more-porous membranes, less material is removed and that removed is typically of higher molecular weight.

7.1 Some Applications

Applications include concentrating milk, treating wastewater for reuse, concentrating stick water from rendering, clarifying juices, sterilizing beer, and recovering

J.P. Clark, *Case Studies in Food Engineering*, Food Engineering Series,
DOI 10.1007/978-1-4419-0420-1_7, © Springer Science+Business Media, LLC 2009

value from wastes. (Stick water is the aqueous phase from centrifuges separating rendered fat and protein from solubles. It contains soluble protein and salts.)

Cheese whey, in particular, is often processed with membranes. Whey is the dilute fluid remaining when protein and fat are coagulated and precipitated from milk as the first step in making cheese. Whey contains soluble protein, lactose (milk sugar), and dissolved salts. An interesting process based on whey makes other valuable food ingredients. Often, whey is concentrated using ultrafiltration to increase the protein content. Whey protein concentrate is a useful food ingredient. However, another valuable component of whey, lactose, is discarded with the minerals and water after ultrafiltration. It is difficult to concentrate the lactose and remove the minerals from the permeate, as must be done for infant formula and pharmaceutical applications of lactose, which require high purity lactose.

A complex membrane process converts the lactose to L-lactic acid and recovers it, as well as a whey protein concentrate. The steps include ultrafiltration to concentrate whey protein, reverse osmosis/diafiltration to concentrate and demineralize lactose, fermentation using a unique *Lactobacillus* strain which selectively produces L-lactic acid instead of a mixture of L- and D-lactic acid, microfiltration to remove bacterial cells, electrodialysis to further demineralize lactic acid, ion exchange to reduce salts further, carbon column to remove color, and evaporation to 80% solids.

This example illustrates the concept of using as much of a feed stream as possible and raising the value of recovered components by increasing their purity. Apparently, the L-lactic acid produced by the special bacterium is key to justifying all the purification steps.

Ceramic membranes. There are two types of ceramic membranes, with 0.2- and 0.01-μm pores. The larger pores are still sufficient to reject colloidal material and bacteria and can therefore be used to clarify and sterilize beverages. The 0.01-μm pore membranes reject molecules whose molecular weight is above about 250,000 Da, which includes many proteins, micelles, and haze forming particles. (Micelles are agglomerations of lipids stabilized by proteins.) Many bacteria are removed by membranes with pores of 0.45 μm.

One focus is on fruit juices, such as apple and cherry and blends, and drinks based on such juices. The membranes have been challenged with spoilage microorganisms and found to give a 5-log reduction. If the same reduction can be demonstrated for pathogens such as *Escherichia coli*, the technique could be used to satisfy the juice HACCP requirements, described in Chapter 6, and still retain a fresh flavor.

Most juices are pasteurized with heat, which can give a cooked flavor. The juice resulting from membrane sterilization, of course, is clear, which is desired in many juices and juice-based products.

Membrane sterilization is not applied to pulpy juices, such as citrus, though perhaps pulp and serum could be treated separately, with only pulp receiving heat treatment, and recombined. Unless enzymes are inactivated, added pulp may settle quickly, so that approach may be doomed. Nonetheless, the ceramic membranes hold promise. In particular, they can operate at higher temperatures than polymeric membranes and need to be cleaned less frequently. Other applications include cold sterilization of beer and clarification of wine. A typical module holds 120 ft^2 of membrane, used in a cross-flow scheme. One module produces 5–10 gal of permeate/min.

A membrane system can be used to treat clean-in-place (CIP) solutions, so cleaning chemicals can be recovered and reused rather than discarded because of buildup of fats and protein soils. The system is based on a membrane that can tolerate the acids and bases used in CIP systems. Such systems also operate at elevated temperatures, up to 140°F. The system uses a tubular configuration and can run for 22 h between cleaning. It is claimed that the cost can be recovered in less than a year, depending on volume, the value of chemicals, and the value of reduced waste disposal.

It has been common to recover CIP solutions, but usually some fresh chemical was added and some used solution discarded to maintain strength and eliminate suspended soils. By filtering out the soil, less make-up chemical is needed, fewer chemicals are discarded, and loads on waste treatment systems are reduced. Cleaning is typically the major source of waste in plants such as dairies and ice cream manufacturers, which are major users of CIP systems.

Membrane processing is close to being a mature technology, with many routine applications, but new materials, such as ceramics, and new synergies with biotechnology, exemplified by the whey fractionation process, demonstrate that the potential of membranes is far from exhausted.

Other inorganic and organic membranes. Porous stainless steel is coated with titanium dioxide sintered to a type of porous ceramic to make a tubular device. In other applications, a renewable organic membrane may be deposited on the titanium dioxide. The ceramic membrane has been tested in snack processing for fines removal from oil, starch recovery from wastewater, and reuse of cleaning solutions. The combination of ceramic and stainless steel permits high-temperature operation, treating of aggressive and corrosive solutions, and use of strong cleaning solutions.

As with other membrane processes, there is a tradeoff between permeation rates and selectivity, with lower rates for removal of smaller particles. The organic membranes are necessary for high selectivity, but they eventually deteriorate at extreme conditions. However, they are easily recreated as necessary, after thorough cleaning of the substrate. The technology has been used in corn syrup production, beer filtration, and pharmaceutical production.

7.2 Process Arrangements

Membrane filtration can be dead end or cross-flow. In dead-end flow, the feed is perpendicular to the membrane surface, while in cross-flow, the feed flows parallel to the surface. Operation may be batch, fed batch, or continuous. (Fed batch means that fresh feed is periodically added to the feed tank as the level drops.) In all membrane applications there is an accumulation of rejected material at the surface of the membrane. This accumulation includes soluble molecules in reverse osmosis and insoluble material in micro- and ultra-filtration. The soluble molecules increase the local osmotic pressure, which then reduces the flux of water through the membrane. Insoluble material increases the resistance to flow of permeate. The effect in each case is that flux rate declines as concentration of the retentate increases.

The effect of concentration on flux rate is best determined by experiment. Theory suggests that flux should be proportional to the negative logarithm of the retained concentration (Mannapperuma 1997). Flux is usually expressed as liters per square meter per hour (lmh) or as gallons per square foot per day (gfd). Because so much of the rate determining resistance occurs in the fluid phase at the surface of the membrane, cross-flow with high velocities of fluid are usually maintained. Because total production rate is proportional to membrane area, it is important to provide as much membrane area in as little volume as possible. This is one appeal of hollow fiber and hollow fine fiber configurations. On the other hand, modules with narrow clearances are prone to plugging by suspended matter.

The simplest configuration is straight batch with total recycle, shown in Fig. 7.1a. A variation is batch with partial recycle (Fig. 7.1b). Partial recycle requires two pumps, one to maintain the velocity past the membrane and one to provide feed

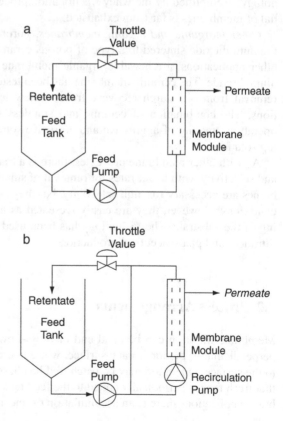

Fig. 7.1 Batch processing systems (**a**) total recycle and (**b**) partial recycle

to compensate for the removal of permeate, while total recycle requires one pump, which does both functions. In fed-batch operation, fresh feed is added to the feed tank to maintain the level. This has the effect of keeping feed concentration low for a while until the supply of feed is depleted.

In different applications either the permeate or the retentate may be the desired product. Reverse osmosis is used to purify water for beverages and laboratory use, but it may also be used to concentrate clarified fruit or vegetable juices by removing water. Such juices are first filtered to remove suspended solids that would foul the very tight reverse osmosis membranes.

To make a membrane process continuous requires a feed and bleed system as shown in Fig. 7.2. Because the membrane sees the highest concentration, flux rate in a single loop feed and bleed system is low, thus requiring the most membrane area for a given task. A multi loop system as seen in Fig. 7.2b has a reduced area requirement. In this approach, each membrane module sees a different feed concentration.

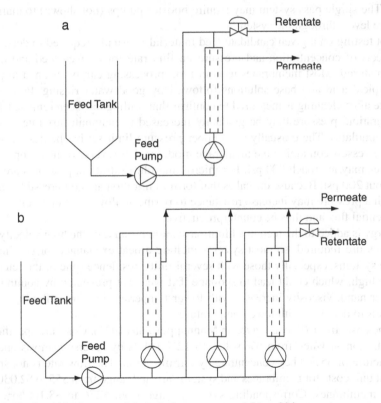

Fig. 7.2 Feed and bleed systems (**a**) single-loop system and (**b**) multi-loop system

A single pass approach is shown in Fig. 7.3, also sometimes called a Christmas tree arrangement because it uses reduced membrane area as the volume of the feed is reduced so that feed velocity past the membrane surface can be maintained at a high

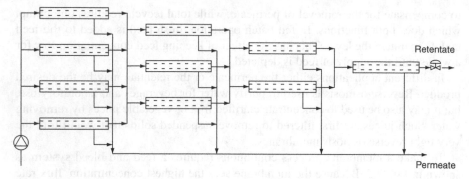

Fig. 7.3 Single pass system

value. The single pass system may require booster pumps (not shown) to maintain pressure levels through the system.

Pilot testing on a given candidate feed material is usually required to determine the effects of concentration and pressure on flux rate for a given feed and membrane material. Most membranes used in food processing can be cleaned in place with typical acid and base solutions followed by good water rinsing. Pure water flux rate after cleaning is measured to confirm that fouling has been removed. During operation, pressure may be gradually increased to maintain flux rate as fouling accumulates. There usually is an upper pressure limit set by the design of the pressure vessels containing the membrane modules. For reverse osmosis, operating pressures may approach 900 psi. For micro- and ultra-filtration, pressures are usually about 200 psi. Because the cakes that form in filtration are compressible, use of too high a pressure may increase resistance to permeate flow more than it increases the potential flux and thus be counterproductive.

Energy is added to the feed in a filtration system because of the high velocity and high pressure required, so most systems include a heat exchanger for cooling. In recycle systems, especially those with several units, residence time of the retentate may be high, which could lead to spoilage if it were not prevented by cooling. On the other hand, viscosity increases with lower temperature for most fluids, so flux rate tends to decrease with lower temperature.

Some costs from 1996 are given in Mannapperuma (1997). Costs are specific for concentration of whey from 6% solids to 12% solids by reverse osmosis and for manufacture of 35% whey concentrate by ultrafiltration. The flow rate is not specified, but unit costs for membranes and systems are and range from $55 to $2,030/m^2 for the membranes. Corresponding systems range from $276 to $8,100/m^2. The ranges are wide because there are large differences in the flux rate, durability, and tolerance for suspended matter. Spiral wound modules are cost effective, but require pre-filtration. A ceramic microfiltration unit was the most expensive, but is also very durable. A more recent quotation, 2007, for a reverse osmosis unit to concentrate vegetable juices was about $1,700/m^2.

7.3 Examples

1. Pilot plant data for reverse osmosis of a food using a spiral wound polymer membrane gave the following relationship for flux in lmh vs. solids %,

$$[F = -1.1B + 43.2] \tag{7.1}$$

    where

    F is flux in liters of permeate per square meter per hour
    B is solids content of feed, %.

    It is desired to treat 69,000 l in 20 h, concentrating from 12 to 28%. Calculate the total permeate assuming perfect rejection of solids. Calculate the area required. Mannapperuma (1997) gives a method using numerical integration to solve for the membrane area required for each increment of concentration and volume change. It is easy to write a simple spreadsheet program to make this calculation. The answer I got is about 88 m^2. A manufacturer quoted a system with 210 m^2. Consider the discrepancy. What are some possible reasons? One possibility is the desire to sell a larger system for more money, always an influence on vendors. Another is a generous allowance for fouling and performance deterioration over time. Still another is the possibility of basing the design on additional data not presented here. (Upon checking, that does not seem to be the case.)

2. A microfiltration system operates in batch mode to clarify a cloudy juice, so the permeate is the product. There are several hundred gallons of retentate remaining at the end of a run, from a batch of 10,000 to 15,000 gal. What, if anything, can be done with this material to avoid disposing of it? The filter operates at about 100°F. Some possibilities include the following: cooling it and allowing it to settle, then decanting some clear juice; centrifuging it; adding it to products that are compatible; using it to make another product; perhaps other ideas. Discuss the pros and cons of each. Some issues include the possibility of spoilage, the need for additional equipment, and the absence of appropriate uses.

3. Think of other possibilities for membrane separations. Mannapperuma (1997) lists the following:

    • Dairy
    • Fruit and vegetable juice
    • Sugar
    • Corn sweetener
    • Wine and brewery
    • Animal products (blood)
    • Process effluents.

    Discuss what specifically might be separated in these cases. Consider other possibilities, for instance, eggs.

7.4 Lessons

1. Membranes offer a low-cost option for purifying water, concentrating fluid foods, and fractionating foods.
2. The process arrangement affects performance and cost.
3. Pilot tests are almost always necessary in real cases because physical properties are usually unknown and variable, though theories can accurately predict the influence of process variables.
4. Designs from equipment vendors may need to be viewed skeptically because their self-interest is not necessarily promoted by offering optimally sized units.
5. The actual membrane module is only part of the process; additional equipment is needed to heat, cool, store, and pretreat.

Chapter 8
Freeze Drying

Freeze drying enjoyed a recent burst of popularity due to the success of breakfast cereals containing freeze-dried berries, such as strawberries, raspberries, and blueberries. Every major cereal manufacturer either had such products or was about to introduce them. According to the Wall Street Journal of May 15, 2003, fruit-containing cereals were the most successful product innovation since sugar-added cereals in the early 1950s.

The recent success of freeze-dried fruit was due, in part, to better packaging of the cereal. Freeze-dried strawberries were introduced in breakfast cereal in the 1960s, but they were very expensive and gradually became soggy as they picked up moisture. Improved packaging helps prevent that moisture pickup. Also, cereals in general have become more expensive, and the manufacturers are making smaller packages, so the berries have less time after a package is opened to pick up moisture.

There are also freeze-dried entrees aimed at the backpacking and survival markets. The products are freeze dried as complete products, rather than mixed from components, because it is believed that this approach retains better flavor and rehydration characteristics.

8.1 Freeze Drying Basics

Freeze drying, or lyophilization, is a preservation process most often associated with coffee, pharmaceuticals, food for space travel, and food for backpackers. It yields a high-quality, lightweight, and easily rehydrated product that retains the original shape of the starting material, unlike conventional drying, in which shrinking and surface hardening can occur.

Freeze drying uses sublimation, the direct conversion of ice to vapor, to remove water from the starting material. Sublimation of water ice occurs below the triple point, which is about 4.5 mmHg water vapor pressure and about 0°C. At or below these conditions, ice becomes vapor without passing through a liquid state. The solids retain their shape, and the vaporizing ice leaves open pores behind. In conventional drying, the water is liquid and the solids are soft and collapse upon removal of the water, causing shrinkage. Soluble solids also can migrate to the surface with water and create impervious surface layers, or case hardening, which impedes rehydration.

Because freeze drying takes place at low temperatures, volatile flavors are retained better than in conventional drying, as is color. Freeze-dried fruits look much like the fresh item and rehydrate quickly in milk or in the mouth, releasing a surprising burst of flavor.

Freeze drying can occur at atmospheric pressure and used to be commonly observed when housewives put wet laundry on outside lines during winter. Typically, the wet clothes would eventually dry but be stiff, as if still frozen. The low humidity of a cold day could fall below the triple point, and the sun would provide the energy needed for sublimation.

Ernest Hemingway once wrote about a freeze-dried tiger carcass found high on an African mountainside, and freeze drying is still used to preserve biological specimens and even whole animals. Freeze-dried microbial cultures can remain viable, and freeze-dried proteins are used as therapeutic and diagnostic agents (Goldblith et al. 1975, Mellor 1978).

Normally, freeze drying occurs under vacuum in an attempt to enhance mass transfer of moisture from the drying material. Often, the limiting rate is actually that of heat transfer. External heat transfer is usually limited by the surface temperature of the material – if this is too high, local melting may occur or there may be discoloration. As material dries, the ice front shrinks away from the surface, leaving a porous and insulating layer of dried material. This becomes the limiting resistance for internal heat transfer.

Research at the University of California at Berkeley in the late 1960s and early 1970s (King 1970, Clark and King 1968) showed that the presence of gas such as nitrogen or helium in the pores of the dried layer could enhance heat transfer in some materials. The low level of gas increased the apparent thermal conductivity of the porous layer, compared to that seen when the pores were completely empty. Too much gas could impede water vapor transfer, so there is an optimum total pressure for freeze drying, which depends on the material but often is higher than the pressure used in normal practice.

The energy of sublimation is a little more than the heat of vaporization of water and can be supplied to the surface by radiation or conduction. In the low pressures usually used, convective heat transfer is not very effective because there is essentially no atmosphere.

Water must be removed by exhausting the vapor with a vacuum pump or ejector or by condensing the vapor to ice. The condenser temperature must be below the ice temperature in the drying material for water to move from the material. Noncondensable gases (residual air) are removed by vacuum pump or ejector.

8.2 Equipment Innovations

One of the major suppliers of commercial freeze-drying equipment offers both batch and continuous freeze dryers with several novel features.

All of the company's dryers use radiant heating from hollow metal platens (shelves) heated with hot water. Not having direct contact with the heating surface provides more uniform heating of the product, compared to trays sitting directly on platens. Trays almost always have irregularities, which makes direct contact non-uniform; thus, the temperature of the platen must be limited to prevent melting or scorching. Many batch freeze dryers utilize trolleys to insert and remove trays of product from the cabinet. This minimizes product handling and equipment cleanup labor.

The larger batch dryers feature a Continuous De-Icing (CDI) system, which utilizes two condensers, cooled by ammonia refrigerant, within the cabinet. One condenser is in communication with the product, while the second is closed off and slightly warmed to melt the accumulated ice. The melting condenser is only slightly elevated in pressure over the operating chamber. The low pressure is regenerated by cooling the condenser prior to its being put back into service.

A similar system called dry condensing can be used in edible oil refining to condense evolved steam and free fatty acids to a more concentrated waste stream than that produced by steam stripping with steam eductors.

Continuous freeze-drying systems use airlocks to introduce and remove trays of material to and from the vacuum chamber. The trays are pushed along tracks between heated platens. The CDI system is also used. Most of the units around the world (none in North America) are used to freeze dry coffee. Concentrated coffee is prepared, frozen, granulated, sorted by size, and then placed onto trays for drying. Equipment capacity is quoted in water removal capability and ranges from 3,300 to 16,000 kg/24 h. Corresponding costs are $4–$10 million, excluding the refrigeration system.

8.3 Other Freeze-Dried Materials

Herbs, vegetables, and fruits are freeze dried as food ingredients. Examples are chives, parsley, and basil.

A recent development is high-ORAC blends for health foods. ORAC, an acronym for *o*xygen *r*adical *a*bsorbance *c*apacity, refers to the ability of foods or nutrients to protect the human body against free radical damage and to the antioxidant power of foods such as fruits and vegetables.

Some other innovative foods are freeze-dried salami as a crisp snack and infused nutraceuticals in fruits and vegetables. Compressed freeze-dried vegetables and fruits have been components of military rations. In this process, products are freeze dried, then slightly moistened to make them flexible, compressed to reduce density, and dried again. It is not an inexpensive process.

A novel continuous freeze dryer, developed by a coffee company, had a chamber 13 ft in diameter and 70 ft long. It used a vibratory conveyer to move frozen coffee granules under radiant heaters. Unfortunately, it had a tendency to make a fine coffee powder, by breaking the granules, that was indistinguishable from that made by spray drying at much lesser cost, so it was scrapped.

Domestic freeze dryers face competition from international sources, as do many other food segments. Because many freeze dryers are batch operations, there is labor involved in loading and unloading, as well as in preparation and packaging. Regions with low labor cost thus have some advantage. On the other hand, freeze-dried items have low bulk density and so incur high shipping costs. Also, precisely because freeze drying is such a good preservation process, it is best to use the highest-quality raw material possible. Freeze-dried materials are very hygroscopic because of their low water content and must be packaged with care.

8.4 Examples

1. My Ph.D. thesis resulted in a US Patent (King and Clark 1969), which synthesized results from other researchers in our group on the thermal and diffusive properties of turkey meat in freeze-drying conditions. (The research was sponsored by the US Department of Agriculture from a budget directed at better utilization of poultry.) There were several concepts to the invention:

   - Operate at a slightly elevated absolute pressure to optimize internal heat transfer
   - Use molecular sieves as a moisture removal means
   - Circulate an inert gas, such as helium or nitrogen, to transfer heat to the surface of the meat pieces
   - Use the heat of adsorption of water vapor on the desiccant to supply the heat of sublimation
   - Choose the sizes of meat pieces and desiccant so they could easily be separated.

   Molecular sieves are synthetic minerals with precisely sized pores and large surface areas, which selectively adsorb water and also find use as catalysts in chemical reactions. They have higher moisture adsorption capacity than many other desiccants, such as silicon dioxide.

   My experiments demonstrated that the process worked and that our relatively simple mathematical model accurately predicted the results. The patent was the first issued to people in that academic department for graduate research; many more, of course, have been issued since. So far as I knew, no one ever applied the invention in practice, though I did hear that it may have been used in Japan.

   Consider why a patent assigned to the US government might not be used industrially. What might have been some good applications for the invention? Hint: we learned later that some classified research had developed a similar process in which frozen droplets of suspended bacterial cultures were mixed with silica gel under vacuum to yield a viable bacterial powder with some potential for use in biological weapons.

What might be some barriers to application? Hint: the freezing points of foods depend on the soluble solids content with the result that fruits, having high sugar content, freeze at lower temperatures than do meats.

2. Write the equations for heat transfer from the environment to the surface of the food, for heat transfer from the surface to the retreating ice front, and for the corresponding transfer of water vapor through the dry layer and then to the condenser, or moisture sink. What are some of the simplifying assumptions that can make these equations relatively easy to solve? (It is easy to make them hard to solve!) Here are some of the assumptions that have actually been used.

   - Simple geometry (infinite slab, cylinder, sphere)
   - No adsorption of water vapor on dried layer surfaces
   - Uniform transport properties (properties actually are dependent on direction of the grain in meat)
   - Uniformly retreating ice front
   - Equilibrium at the ice front (Water vapor pressure is pure component vapor pressure at ice temperature. The Clausius-Clapyron Equation correlates water vapor pressure with the reciprocal of absolute temperature.)
   - Thermal conductivity of frozen core is so much larger than that of the dried layer that the core can be assumed to be at a uniform temperature.

   If you can, use the equations to estimate drying times for a range of foods. You may need to consult the reference literature for relevant physical properties. (Giving the reader a motivation to learn how to find physical properties of foods is one reason they are not provided here. This is a necessary skill for the food engineer.)

3. In the freeze drying of pharmaceuticals, such as purified therapeutic proteins, the normal procedure is to deposit a precise amount of solution or suspension in a glass vial and then loosely place a rubber stopper in the top of the vial. The vials are placed in trays and the trays placed on hollow shelves through which a cold fluid circulates to freeze the contents. The chamber is closed, a vacuum is drawn and a hot fluid is circulated through the shelves. Evolved water vapor is removed by a vacuum pump past a refrigerated condenser. When the drying is complete, the shelves are moved closer together, forcing the rubber stopper closed, the vacuum is relieved, the chamber is opened, and the trays removed.

   Consider some of the issues that apply to this application. The filling is done in a clean room. What is a clean room? How is it maintained? (A clean room is classified by the number of fine particles suspended in a cubic meter of air – 1, 10, 100, 1,000, etc. This is achieved by maintaining laminar flow of air from the ceiling to the floor and passing all air through high-efficiency filters (HEPA filters).) It is expensive to circulate the large volumes of air and to maintain the filters. How would you minimize these costs? The major resistance in freeze drying under these conditions is the passage of water vapor past the loosely fitted rubber stopper. What can be done about that? How much ice will build up in a typical run, say from 5,000 vials each holding 20 ml of a 5% solution? How big should the condenser be? It is common for each batch to take 48 h to dry. What is the motivation, if any, to optimize freeze drying times of these often very valuable materials?

4. Consider again the production of entrees for backpackers, boaters, and military rations. These are dishes such as beef stew, chicken and rice, shrimp curry, and chili. The usual practice is to formulate these in a kettle, fill into trays, freeze, either in the chamber or separately, and then freeze dry under vacuum, typically overnight. The dry slab is broken up into small pieces and these are filled into pouches and sealed. Alternatively, the slabs can be cut into portion-sized pieces and then these are packaged. Consider some of the issues that might, and probably do, arise in this process. Examples are as follows:

- Production of fines (How can you minimize them? What can you do with them?)
- Different drying rates of large and small pieces and of different materials
- Handling of fragile slabs and cut pieces
- Size reduction of slabs
- Getting slabs out of trays
- Consistency in filling of solids.

There are many others you can probably imagine. Discuss the benefits and challenges of another approach, in which components, such as meat pieces, sauce droplets, and vegetables, would be separately freeze dried and then combined, perhaps by a multi-shot fill into each pouch. (Multi-shot fill means that several different components are deposited separately into a container rather than being mixed and then filled.)

In this approach, freeze-drying conditions for each component could be separately optimized. However, that would require a chamber for each or a provision for storing the ingredients until all were ready. Mixing separate components with widely varying particle sizes is sure to experience segregation and, since the pieces are fragile, there will be breakage and fines generation. Multi-shot filling avoids some of these issues but requires additional capital in the form of additional fillers.

8.5 Lessons

1. Freeze drying is more expensive than conventional drying, but it gives a superior product.
2. Compared to conventional drying, it is relatively easy to model mathematically, and the insights from doing so can often be transferred to other applications.
3. Both mixtures and individual components are freeze dried, and there are good reasons for each.

4. Multi-shot filling can be a solution to the packaging of mixtures with a wide range of particle sizes.
5. Because of the high value of some freeze-dried products, there is rarely any great concern over reducing cycle times to save operating costs. It is more important to increase yield, prevent losses, and preserve quality.

Part II
Processes Based on Biochemical Reactions and Thermal Treatment

Chapter 9
Continuous Thermal Processing

The basics of thermal processing are discussed at great length in many other texts (Singh and Heldman 2001, Ibarz and Barbosa-Canovas 2003, Karel and Lund 2003, Charm 1971, among many others), so they will only be summarized here. For a given food and for a given objective (cooking, refrigerated shelf life, ambient shelf life, complete sterility), there is a time at a specified temperature that must be achieved. In a continuous process this hold time is controlled by the length of the hold tube for a specified flow rate. It is important to know the flow regime in the hold tube – whether it is laminar or turbulent – as this determines the velocity profile and thus the hold time. Many fluid foods are viscous and non-Newtonian in their flow behavior and so estimating heat transfer coefficients and pressure drops can be challenging.

The same basic flow diagram shown in Fig. 9.1 applies to many different processes, including pasteurization, cooking, hot filling, and aseptic processing. Aseptic processing will be discussed in some detail because it is the most complex and illustrates many of the important lessons.

9.1 Aseptic Processing

Aseptic processing involves sterilizing a fluid food and its containers separately, cooling the food, and then filling the sterile food into the sterile containers in a sterile atmosphere (David et al. 1996). The general process is very flexible, with containers ranging from 8 oz paperboard boxes up to steel tanks holding 1 million gallons. Products treated by the process include fruit juices, dairy products, purees, soups, and cheese sauces. These will be discussed briefly.

9.1.1 New Developments in Aseptic Processing

One new development in aseptic processing is low-acid beverages in plastic bottles, such as flavored teas. Teas are often acidified so that they can be treated with a milder process than is required at their normal pH of 4.8–5.6. (Products with pH

J.P. Clark, *Case Studies in Food Engineering*, Food Engineering Series,
DOI 10.1007/978-1-4419-0420-1_9, © Springer Science+Business Media, LLC 2009

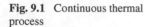

Fig. 9.1 Continuous thermal process

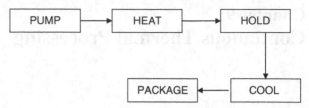

below 4.6 are considered high acid and can be commercially sterilized at temperatures below the atmospheric boiling point of water, whereas low-acid foods must be sterilized at elevated pressures and temperatures – closer to 250°F. The reason is that *Clostridium botulinum* does not produce its deadly toxin in acid conditions. Most other pathogens and spoilage microorganisms are killed by any treatment that kills *C. botulinum* spores. It has recently been observed that acid-tolerant pathogens such as *Escherichia coli* O157:H7, *Listeria monocytogenes*, and *Salmonella* can survive in high-acid products. The pathogens are reduced with time of exposure to pH below 4.0.) Teas are believed to taste better near their natural pH, but these higher pH values make the beverages low-acid foods, requiring more rigorous sterilization processes and attracting more stringent scrutiny of the filling equipment by the Food and Drug Administration.

Two manufacturers now have linear filling machines for plastic containers accepted by FDA. Linear machines operate at relatively low filling rates of 200–300 bottles/min. A higher-speed technology is rotary filling, often used in Europe for fillers considered to be aseptic there but not yet accepted by FDA. The issues are the use in Europe of a sanitizer composed of peracetic acid and hydrogen peroxide and the concern that in a rotary filler, air flow is turbulent, not laminar. So far as is known, no rotary filling machines have yet been accepted as aseptic by US regulatory agencies.

The objection to the chemical blend stems from some published research that asserts that the mixture is a sanitizer, not a sterilant. Straight (30%) hydrogen peroxide and heat is used in the United States for sterilizing aseptic containers and closures. However, hydrogen peroxide seems to have some interaction with polyethylene terephthalate (PET), a desirable material of construction for bottles. Current bottles used for aseptic filling are usually high-density polyethylene (HDPE), but it does not have the same clarity as PET.

9.1.2 Aseptic Processing Basics

Some of the issues alluded to earlier may need to be clarified. Aseptic processing is the separate sterilization of pumpable food, containers, and closures, followed by the cooling of the food and the filling and sealing of the containers in a sterile atmosphere. Once the container is hermetically sealed, it can leave the aseptic filling zone for further operations, such as labeling, cartoning, and case packing.

Food. Sterilization of the food is normally achieved by heating, often in scraped-surface heat exchangers, tubular heat exchangers, or by direct steam injection. After heating, the food is held for a specified time, usually by flowing through a holding tube, and then cooled. Viscous foods may be in laminar flow in a holding tube. Since many foods are non-Newtonian fluids, the velocity profile, which determines the residence time distribution in the tube, may be flatter than for a Newtonian fluid in laminar flow. When particles are present, there is concern about the temperature profile that may exist between the fluid and the center of the particle.

Typically, it takes about twice as much heat-exchange surface for cooling as it does for heating, depending on the desired final temperature. Scraped-surface heat exchangers are efficient, but can incur high maintenance costs, as they have bearings and wearing surfaces. Tubular heat exchangers have fewer moving parts. A common design is a triple tube in which food flows through an annular space between a center tube and an outer tube, both of which contain heating or cooling media. Hot water or steam is used for heating and chilled water or glycol solutions for cooling.

Containers. Sterilization of containers may be by heat or chemicals. Metal cans are sterilized with steam. Thermoformed containers are sterilized by the heat used to soften and deform a sheet of plastic to make the container. Paperboard or preformed containers are sterilized by chemicals – hydrogen peroxide in the United States. Heat is used to dry and decompose the residual hydrogen peroxide.

Closures. Closures may be metal lids for cans, sterilized by steam; or heat sealed films for thermoformed packages or bottles, sterilized with chemicals as roll stock that may be perforated and from which the closure is cut or pushed.

Bulk containers. A special case of container is bulk containers, which may be composite bags or rigid tanks. The bulk bags are sterilized by irradiation, while the bulk tanks are sterilized by chemicals. The tanks are filled with a chemical solution, which is then displaced with sterile nitrogen. The nitrogen is sterilized by filtration.

Filling system. The filling environment is sterilized by heat or chemicals and then maintained by sterile filtration of air or inert gas. The filters themselves must be validated and sterilized, lest they become a source of contamination. Typically there are two filters in series in case one fails.

All fittings, valves, and penetrations of the filling area must be designed for ease of cleaning and for the ability to be sterilized in place. One of the reasons there are relatively few accepted aseptic fillers is the challenge of meeting the stringent requirements for FDA acceptance.

9.1.3 Regulation of Aseptic Processes

FDA does not "approve" aseptic processes or equipment. Rather, it accepts a filing detailing the design, process conditions, tests, and control procedures. Often, FDA raises questions or suggests additional tests until it is satisfied. At that point, FDA sends a letter saying the filing is accepted.

There was a cooperative effort to demonstrate how to verify a thermal process for a model food containing larger particles, but so far it does not appear that any commercial effort in the United States has been made to do so. Food particles smaller than 0.25 in. in diameter or on edge are considered homogeneous. Existing aseptically processed foods with particles include tapioca pudding, oatmeal, and a cheese sauce with jalapeño pieces.

9.1.4 Aseptically Processed Foods

Some foods that are aseptically processed include the following.

Baby foods. Baby foods manufacturers may be some of the largest users of the technology. Most, if not all, of the traditional small glass jars are being replaced with plastic containers. Most fruits and some vegetables are already on the shelves in the new, thermoformed packages. Fruit juices have been in hot-filled plastic bottles for several years.

Baby food purees come in three main serving sizes (2, 4, and 6 oz), and some differ in particle size as well, for the different eating skill levels of different-aged babies. Fruits tend to be high-acid foods, while vegetables and meats are low-acid foods. Viscosity is also a factor in thermal processing and in filling.

Baby foods, compared to other foods that might be aseptically processed, have the advantage that they are inherently homogeneous and so do not pose the complexities of determining thermal processes for foods with larger particles.

Cheese sauces and puddings. These are some of the older products to be successfully aseptically processed. Before plastic packaging became widely used, cheese sauces were processed for food service in No. 10 cans that had been sterilized with superheated steam and filled in a steam-sterilized environment. Lids were likewise sterilized with steam and seamed in the same environment. Puddings have been filled into small cans, but now are most often found in thermoformed polymer containers.

Tomato paste. Tomato paste has been aseptically processed and filled into 55-gal drums for some time. The advantage is the improvement in quality from cooling externally rather than trying to cool a large container after hot filling.

9.1.5 Typical Process Equipment

Aseptic processing of viscous foods often uses corrugated tubular and spiral-coil heat exchangers. The corrugation refers to embossments of the surface of the tubes to enhance the heat-transfer efficiency. Equipment companies often fabricate complete process systems, including all types of tanks, pumps, valves, and necessary controls for the process requirements. Such equipment is used for purees, tomato products, coffee creamers, whipping cream, and other viscous foods.

9.1.6 Foodservice Opportunities

A very promising area for aseptic processing is foodservice, for semi-solid foods such as soups, sauces, cheese sauces, puddings, ice cream mix, and purees, especially in quick-serve and family-style restaurants, where refrigerated storage is in short supply. Typical packages are 5- to 10-lb bags in boxes. Critical issues are filling and dispensing fitments (the special valves used to remove just the right quantity of the contents as needed), increasing processing rates, and maintaining product integrity and safety of low-acid foods at the point of sale. The success of bag-in-box wine is a good example of the potential.

9.1.7 Bulk Aseptic Storage

Philip E. Nelson, Professor and retired chair of the Department of Food Science at Purdue University, West Lafayette, IN, has been responsible for major developments in bulk aseptic storage of foods, and was awarded the World Food Prize in 2007 for his achievements. Early in his career, he was interested in prolonging the tomato processing season in the Midwest because of his own background with a family tomato business. A fabricator of tanks for breweries agreed to help.

Nelson and his group over a number of years developed valves, filters, and procedures to enable the storage of high-acid foods, such as tomatoes, in bulk tanks. A key discovery was the need to flood the tanks with a sterilant solution of iodophor. Spraying, as in cleaning, did not work because it did not sterilize the atmosphere. After the solution was drained, it was replaced with sterile nitrogen.

The iodophor was used at no more than 25 parts per million (ppm), so that it did not need to be rinsed. Purdue researchers found that an epoxy lining was superior to even stainless steel, which also meant that tanks could be made of epoxy-coated carbon steel and therefore be less expensive than stainless steel construction.

In parallel to the research on bulk tanks, work at Purdue helped to develop the bulk bag system. The original bags were made to store battery acid, but with the application of irradiation sterilization and design of sterile fitments to fill and empty, the bags could be used to store fruits and vegetables for later processing. Now it is common to store purees and concentrates of apples, peaches, pears, and mangoes and later repackage them as such or in other products.

Some low-acid foods are stored in bags, but not in bulk. Citrus juices are stored in very large tanks, up to 1.7 million gallons, but must be refrigerated because they are subject to browning. Seagoing tankers with 16 half-million-gallon tanks routinely take not-from-concentrate orange juice from Brazil to Europe and the United States.

9.2 Other Applications of the Flow Sheet

9.2.1 Cooking

Baby foods were historically made by cooking whole fruits and vegetables by direct contact with steam, over long periods of time, and then forcing the softened material through screens, leaving skin, seeds, and stems behind. Nutrients were lost to the discarded condensate, overcooked flavors developed, and the bitter stems and seeds contributed undesirable flavors. A major improvement in nutritive value and flavor was made by the use of cold extraction, the reduction to a pulp or puree of the raw fruit or vegetable, followed by brief cooking to soften, and, finally, screening of the hot pulp. Each commodity requires some unique preparation step. Peaches have their pit removed; squash (which look like good sized pumpkins) need to be chopped into smaller pieces and have their seeds removed; carrots and apples are washed; and frozen raw materials need to be thawed and chopped. The screens and production rates are tailored to each variety. Then the puree passes through a heat exchanger and hold tube very similar to those of an aseptic system. The puree is not cooled, but rather is forced through a smaller screen and then is pumped to filling. Some products are filled into glass jars and retorted while others go through an aseptic system and are either bulk filled into large bags or into retail packages.

Most agricultural commodities are seasonal, meaning an entire year's supply must be processed in a relatively short time period. The ability to store in bulk bags means that decisions about specific package sizes and formulations can be deferred until market demand is clear.

The improved cooking process, using the continuous thermal process, resulted in a significant increase in sales for the company that commercialized it.

9.2.2 Pasteurization

Pasteurization is distinguished from sterilization in using milder conditions with the intent to remove all pathogens and some, but not all, spoilage microorganisms. Milder conditions are used to protect flavor, color, and nutrients, which are easily damaged by high temperatures. By law, raw milk must be pasteurized using specified conditions of time and temperature. Raw milk frequently contains *Salmonella* and other harmful bacteria, which are destroyed relatively easily under the specified conditions. Table 9.1 gives the values found in the Pasteurized Milk Order (PMO), followed by most states.

The values illustrate that time required for a given amount of reduction in microorganisms decline exponentially as temperature increases. Other reactions occur as foods such as milk are heated, many of which are undesirable, including darkening of color, caramelization of sugars, and reactions between sugars and amino acids that lead to colors and flavors. The last reaction mentioned is the

Table 9.1 US federal pasteurized milk order (PMO) temperature vs. time

Temperature (°C)	Time (s)
63	1,800.00
72	15.00
89	1.00
90	0.50
94	0.10
96	0.05
100	0.01

Fig. 9.2 Pasteurization with regeneration

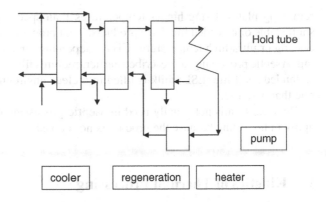

Hold tube

pump

cooler   regeneration   heater

browning or Maillard Reaction and can be beneficial, as in grilling meat. The impact of temperature on many undesirable reactions, including vitamin degradation and cooking, is less than it is on the destruction of microorganisms, so it is usually wise to operate at the highest temperature and shortest time that the equipment and food can tolerate.

Thermal processing consumes energy for both heating and cooling, so it is logical to consider ways to reduce this cost. One that is common in pasteurization is known as regeneration, as illustrated in Fig. 9.2. The concept is to use the hot stream to pre-heat the cool raw feed while cooling the hot stream. Because all heat transfer requires a finite temperature difference between the streams exchanging heat, there still needs to be a final heating step and a final cooling step, using outside sources, such as hot water or steam and chilled glycol or ammonia. Nonetheless, a well-designed pasteurizer with regeneration can recover around 90% of the energy in the hot stream, which represents a significant savings.

As with most process modifications to improve efficiency, there is an increase in complexity. First, it is necessary to establish a specific pressure profile in the flowing product stream so that the treated stream is at a higher pressure than the raw stream. This is so that if there is a leak through the wall of the heat exchanger, the flow will be from "clean" to "dirty" avoiding contamination. However, since pressure falls in the direction of flow, there needs to be a booster pump on the product stream to raise its pressure. Usually, this is placed on the hot stream as it exits the regeneration section and before it goes through the final heater and hold tube, but it can be placed after the hold tube, as shown in Fig. 9.2. There also needs to be a diversion valve at the end of the cooling section, which is controlled by the temperature at the end of the hold tube. If that temperature drops below the set point, the stream is directed back to the feed tank rather than going forward as safe product. The diversion occurs after cooling so the feed temperature is not distorted by adding a hot stream.

In dairies and some other industries, the booster pump may also be a homogenizer because homogenization is improved by elevated temperature. The process as described is called high temperature short time (HTST). Often, the heat exchangers are plates mounted on a common frame. This makes for a compact equipment arrangement and the configuration can be changed relatively easily by adding or removing plates. Using higher temperatures than specified in Table 9.1 can give shelf stable or extended life dairy products. That process is called either aseptic processing or ultra-high temperature (UHT), depending on the conditions and packaging. Aseptic processing, as described earlier, requires filling in a sterile environment. Extended shelf life (ESL) milk is filled in a clean environment, but not necessarily one that is sterile.

Regeneration is not usually used in aseptic processing or cooking and does not apply in hot filling because the product is not cooled.

9.3 Kinetics of Thermal Processing

The destruction of microorganisms and the inactivation of enzymes is usually a first-order process, meaning that a semi-log plot of the reduction in numbers or activity vs. time is a straight line. See Fig. 9.3. The time to reduce microorganism numbers by a factor of 10 is called the decimal reduction time, D, with units of seconds or minutes. Thermal processes are often expressed as the number of decimal reductions that are intended or achieved, as in $5D$ (or 5 log, meaning the same thing). The final level of contamination after a given process depends on the initial level. When the initial level of contamination is such that after a given number of decimal reductions, the final level is less than one per container, the significance is understood as a probability of spoilage or of potential hazard. A common objective is to achieve less than one chance in 10,000 of a surviving spore in any container.

Confirming such a result is obviously difficult, requiring enormous sample sizes. In practice, thermal processes are often developed by inoculating containers of a specific food with large amounts of a surrogate microbe – one that is safe to handle in the laboratory but has known thermal death time kinetics similar to a pathogen of concern. Studies are designed to have measurable amounts of survivors and then actual processes are specified by calculation.

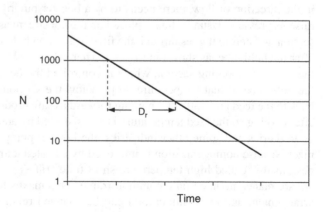

Fig. 9.3 Log N vs. time

9.3.1 The Effect of Temperature

The studies previously described typically are done at constant temperature (isothermal), using small samples and controlled baths, so the time to heat and cool is very short compared with the measured time at the specific temperature. In practice, with real containers or with flowing streams, the food sees a varying temperature history, rising from its initial temperature, holding for some time period, and then cooling over some time period.

In chemical reactions, the first order reaction rate constant, which is the reciprocal of D in this case, is known to vary exponentially with the reciprocal of absolute temperature. However, over relatively short temperature ranges, Log D has been found to vary linearly with temperature. See Fig. 9.4. The slope of this line is represented by z, the number of degrees required to change D by a factor of 10. The z value often used for *C. botulinum* is $10°C$ ($18°F$). In the absence of specific data, this value is often used for other organisms. The equivalent of z for other reactions, such as nutrient loss or color development, is usually larger, meaning that temperature has less of an effect – it takes more degrees to increase the rate – thus it is advantageous to use high temperatures and short times when possible to get an adequate reduction in pathogens and reduce damage from nutrient loss or color development.

Fig. 9.4 TDT vs. temperature

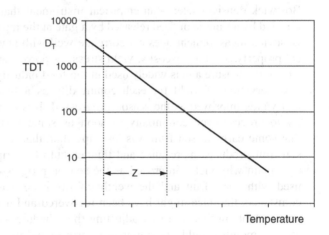

9.4 Examples

1. A baby food company conceived the idea of cold extraction (crushing and straining) fruits and vegetables, rapidly heating the puree, further straining, and then filling jars and retorting. The process looked, in principle, very much like the continuous thermal process discussed in this chapter. As supplied, the process had 24 triple tube heat exchangers in series, two progressing cavity pumps, hold

tubes that could be adjusted by use of jumpers to provide residence times for high- or low-acid foods, a centrifuge to thicken products, and was intended to operate over the range of 15–60 gpm for a wide variety of fruits and vegetables – peaches, apples, pears, squash, carrots, spinach, peas, sweet potatoes, and green beans. Operation was not satisfactory for a number of reasons:

- The pumps' stators failed frequently, contaminating food with bits of black polymer.
- The pump after the hot extractor could not remove product at the highest rate.
- It was hard to achieve desired consistency (thickness or mouth feel) for some products even with the centrifuge.
- Operation of some preparation equipment (destoner for peaches, conveyor for green beans) was inconsistent.

Over a period of about 2 years, the process was modified to address these and other issues. *Before reading further, make a list of possible causes and remedies.* One of the first observations was that there was not just one process – there were as many unique processes as there were products, at least nine and maybe more. Each product had a unique preparation step; each had a specific time and temperature requirement (similar for all high-acid foods and a different combination for all low-acid foods); and each had a specific consistency target (measured by Bostwick consistometer – an empirical instrument that measures the distance traveled by a sample of food released by a gate at the top of a sloping ramp). The Bostwick measurement does not correlate well with any fundamental rheological property, such as viscosity, yield stress, or power law index. However, it is simple to measure and is widely used in the food industry.

Observation of yield for each commodity as a function of time revealed that yields improved as the season progressed. It was customary in the industry to process each commodity as soon as some was available. This meant that some early-season fruit was less ripe than that harvested later. Early season fruit produces more juice and thinner pulp than riper fruit, which helped to explain why yields improved as the season progressed. The centrifuge was used with early fruit and the weight of the juice was lost. (In another circumstance, the juice might have been recovered and used, but that option was not available in this case.) By adjusting the schedule – delaying production by about 1 month – yields were greatly improved and the centrifuge was rarely if ever used.

By observing actual practice over several seasons, it was learned that almost all runs were at a rate of 30 gpm. A turndown ration of 4:1 is relatively rare in most food equipment and is only achieved by designing for a worst case that may hardly ever be encountered. Measuring surface temperatures of the heat exchangers revealed that most of the heating occurred in the first few modules; about 75% of the heat transfer equipment was not doing anything. Further, it was contributing to pressure drop and was extending the hold time beyond that which was intended. (The surface temperatures were measured because there were no instrument wells along the heat exchangers, an oversight that is common as a cost savings but which makes performance studies difficult. The surface temperatures were lower than the fluid temperature because of the resistance of the tube wall, but their relative magnitude was more important than their absolute values.) It was suggested that the heating section be reconfigured to have eight modules instead of 24. Other uses were found for the excess tubes.

As a side observation, it is common for equipment vendors to oversize equipment, reasoning that underperformance is more objectionable than additional cost. Oversizing by a factor of 3 or 4, however, is a bit excessive, in my opinion.

The reason there were two pumps was a belief that much of the pressure drop would occur in the first few heat exchangers, where the puree was cold and most viscous. As it warms up, the viscosity decreases and so does the pressure drop. The failure of the pump stators was caused by a combination of defective polymer and the high temperature and pressure drop encountered. It was suggested that the two pumps be replaced by one rotary lobe positive displacement pump after removing the excess heat exchangers. It turned out that the pressure drop was still higher than the rating for the pump (about 400 psi vs. a limit of 300 psi). The solution was to re-arrange the heat exchangers into two banks of four. This cuts the pressure drop dramatically. *How much?* The pressure drop is proportional to the flow rate and to the pipe length. By dividing the flow in half, the pressure drop was cut in half and by reducing the length, the pressure drop was halved again for a total reduction by a factor of 4. This was confirmed in practice.

The pump supplying the filler was still struggling because the pipe was 2 in. tubing and the length was several hundred feet. A larger diameter pipe should have been used. Going from 2 in. to 3 in. diameter would have reduced the pressure drop by more than half, but at considerable expense. Instead, a second pump was installed about half way down the line, using an available tank for surge. Feed to the supply pump from the second extractor occasionally stopped because of the thickness of the puree. A wide hopper was installed instead of the narrow pipe to maintain a positive suction head at the pump.

Later the company changed from filling glass jars to using an aseptic process with thermoformed plastic containers. This process was installed close to the cooking equipment. The aseptic process looks very much like the cooking process, with triple tube heaters, an adjustable hold tube, and a cooling section. It was sized for one flow rate, which matched that of the cooking equipment. The net effect is that fruits and vegetables that once were cooked for several hours – once in the cooking step and again in the container – were now cooked for a total of about 5 min for low-acid foods and even less for high-acid fruits. There was a dramatic improvement in flavor, which was reflected in an increase in sales and market share.

2. A company that processed vegetable purees as food ingredients had a heat – hold – cool system using scraped surface heat exchangers. What they considered a hold tube was installed between two of the heaters. They had considered all the time the fluid spent in the system as their hold time, using the volume of the heat exchangers. They were experiencing about 30% rejection of product for high

microbial counts. This product was reprocessed, incurring costs and impacting flavor. The company also wanted to increase production.

It was explained to this company that they were not performing an adequate pasteurization process, which was intended to be equivalent to 180°F for 30 s. The hold time did not begin until the product was at target temperature, which was ensured by use of a divert valve triggered by the temperature at the exit of the hold tube. The hold tube needed to be the proper length and needed to be horizontal. Because of the viscosity of the puree, it was determined that the flow was laminar, meaning that the hold tube should be twice as long as it would be to give an average residence time of 30 s. *Calculate the Reynolds number, assuming an effective viscosity of 10,000 cp. For a flow of 22 gpm, how long would a 2 or 3 in. hold tube be?*

It is common practice in aseptic systems and other continuous thermal systems to have twice the cooling heat exchange area as the heating heat exchange area. However, this company wanted to cool to close to 40°F rather than 70–80°F, which is often adequate. This means they needed more than six coolers to match the three heaters they used. It was suggested that they consider using triple tube coolers instead of more scraped surface heat exchangers because triple tubes are less expensive and have lower maintenance costs. Scraped surface heat exchangers are good with very viscous products, but triple tubes are competitive in performance.

Originally, the company used an undersized glycol coolant system and was not getting the exit temperatures they wanted. It was suggested that they use cooling water on three or four coolers and glycol only on the last coolers. While there is a cost for cold water, this water could be used elsewhere in the plant, for washing, and so it provided essentially free cooling.

3. An existing pasteurizer uses plate heat exchangers on a common frame to heat with hot product in a regeneration section, heat with hot water, hold in a tube, partially cool through the regeneration section, and then filter while still warm. The hold tube was sized assuming turbulent flow because the product viscosity was about 10 cp. *Calculate the Reynolds number for a flow rate of 15 gpm. What is missing from the system as described?* The system needed the addition of a booster pump to ensure that the treated product stream was at a higher pressure than the raw feed. It also needed a divert valve. The divert valve is triggered by the temperature at the outlet of the hold tube because sometimes there is a temperature drop down the length of the hold tube. The actual valve is often at the outlet of the cooling section so that the cooled product is diverted to the feed tank. Otherwise, the temperature of the feed might vary, causing the control system to exercise more than it should, and even to encourage spoilage of the feed.

4. A manufacturer of dairy products, such as milkshake mix for food service, complained that new products it developed using its pilot plant aseptic system had difficulties transferring to the large-scale manufacturing plant across the street. Typically products made in the plant were darker in color and more cooked in flavor than the same formula had been in the pilot plant. Upon inquiry at the plant, it was learned that the plant operators routinely increased the target temperature

"just to be safe." Furthermore, they routinely ran the manufacturing system at a lower flow rate than that for which it was designed. Thus, if the specified and developed process called for heating to 280°F and holding for 15 s, the plant might set controls for 295°F and run their 50 gpm system at 45 gpm. *Estimate the impact of these adjustments on the thermal process received by the product, assuming a z value of 18 F.* The impact of temperature is calculated as $10^{(T-T_0)/z}$, where T is actual temperature and T_0 is reference temperature. In this case, the combination of higher temperature and lower flow rate meant that the product received about 7.5 times the thermal treatment that was intended. That is usually enough to cause considerable darkening in a fragile dairy product.

To make matters worse, the plant system was capable of running at even higher rates but had no provision for taking excess heat exchangers out of service. As a result, the required heating was accomplished in the first part of the heaters and the rest served as additional hold tube. Many operators are not aware of this phenomenon and most systems do not have jumpers to permit removal of a portion of the heat exchange area. It would be a considerable improvement if they did.

Operators should not be permitted to deviate from established processes even when they mean to be safe. It is the responsibility of developers to include adequate safety factors and additional measures should not be applied. This requires mutual confidence on the part of research and operations people.

9.5   Lessons

1. The continuous thermal processing system is more complex than it may appear and involves heat transfer, fluid flow patterns, and pressure drop as significant considerations.
2. Continuous thermal systems need to be properly instrumented and have provisions for adaptation if flow rates or required process conditions change. In particular, there should be temperature indicators after every heat transfer module.
3. Hold tube length and amount of heating area should be capable of being changed if flow rate or process conditions change.
4. Parallel flow in heating and cooling can reduce pressure drop with no increase in complexity or cost.
5. Equipment is routinely overdesigned and oversized. There is some economic motivation by vendors to do this, but primarily it is done to accommodate worst-case design specifications. Buyers should carefully examine their requirements and challenge requests for excessive flexibility. In particular, most food equipment operates efficiently over 2:1 range of capacity or rate, but rarely over a wider range.

Chapter 10
Retort Pouch Foods

Retort pouch foods are a specific application of batch thermal processing, in which the basics still apply, but the means of achieving a specified time and temperature treatment are different from those previously described. Developing the first commercial application of the technology in the United States was the occasion for learning some significant lessons about complex machinery. The retort pouch is a good example of an apparently great idea never quite achieving its expected potential. In many ways, it was the right answer to the wrong question.

10.1 History

The retort pouch package was developed at the US Army Natick Laboratory, where rations, uniforms, and other supplies are developed for all the armed forces, as a replacement for the metal can. It is interesting to note that the metal can and the technology of preserving by heating under pressure were developed by Nicholas Appert in France in response to a challenge by Napoleon Bonaparte for a safe and nutritious means of feeding his troops. There is a long history of developments for military purposes finding civilian applications.

There were two major motivations for the retort pouch package. First was a desire for a soft package which would not injure soldiers diving to the ground with packages in their pockets, as often happened while carrying the cans used in C (for canned) rations. The second objective was to improve the sensory quality of the contents by providing a thin profile package and thus reducing the length of the high temperature cook required to ensure shelf stability of low-acid foods.

The major concern in preserving foods in closed containers (hermetically sealed, that is, closed against intrusion of air) for long periods of time at ambient (room) temperature is the prevention of the growth of *Clostridium botulinum*, a spore-forming bacterium that produces a deadly toxin under certain conditions. The conditions for toxin formation are pH above 4.6, adequate water activity, low oxygen concentration (anaerobic conditions), and temperature about 75°F. Foods with pH below 4.6 are naturally acid, made acidic by addition of edible acids, such as vinegar, or made acidic by fermentation, which produces lactic acid, as in fermented

J.P. Clark, *Case Studies in Food Engineering*, Food Engineering Series,
DOI 10.1007/978-1-4419-0420-1_10, © Springer Science+Business Media, LLC 2009

pickles, and do not support the production of toxin. Some other pathogens have been found, surprisingly, to tolerate acidic conditions, but they do not live long in such an environment.

Spores of bacteria are a life form, which are very resistant to heat and other hostile conditions. Spores become vegetative microorganisms when environmental conditions are favorable. Only growing, vegetative microbes produce toxin, but food offers such conditions, so the objective of canning processes is to destroy all spores of *C. botulinum*. The conditions that do this, typically 250°F for several minutes, also destroy almost all spoilage microorganisms and inactivate most enzymes. The resulting food is considered commercially sterile, to distinguish it from absolute sterility. Commercially sterile foods may decline in quality over long periods of time due to slow biochemical reactions, but they remain safe to consume for many years.

The conditions imposed by the military for its canned rations are typically more severe than normal commercial practice because the military expected its canned foods to be exposed to extreme conditions, especially elevated temperatures that can accelerate spoilage, if there are surviving microorganisms. The high temperatures encountered in commercial canning processes have negative effects on the texture, color, and taste of most foods, and the conditions applied to military rations, being more severe, aggravate this situation, giving rise to the poor reputation of C rations in particular and of military food in general. The retort pouch was thought to improve food quality by permitting a shorter process time due to the thin profile of the package.

The time required for a canning process is that necessary to heat the slowest heating point of a container to the target temperature for the required time. This corresponds to the slowest heating portion of fluid in a continuous thermal process, as previously discussed. Heat is transferred within a can by a combination of conduction and convection, depending on the physical properties of the food. Solidly packed meat and thick soups and sauces transfer heat mostly by conduction, while thinner products can develop convection currents due to differences in density as the temperature increases. Thin products also benefit from mechanical agitation, in which containers are tumbled end over end or are rotated in cages or by sliding in a helical track. A new retort design shakes the basket containing cans or jars to promote heat transfer.

The time for heat transfer to reach a given objective is roughly proportional to the square (2nd power) of the characteristic distance over which heat must move – the radius of a cylinder, for instance, or the half thickness of a slab heated from both sides. The geometry of the retort pouch resembles a slab and can have a half thickness much less than the radius of a can holding similar volume. This means that the time to sterilize could be significantly reduced and so the quality of the food should be significantly better. See Fig. 10.1.

Traditional canned foods are packaged in metal cans of various sizes, glass jars, or plastic containers. The contents are filled hot and so after sealing, a vacuum is produced within the container. This is relevant because containers are typically cooled by immersion in water. The vacuum can suck cooling water through small leaks in

Fig. 10.1 Comparison of pouch and can

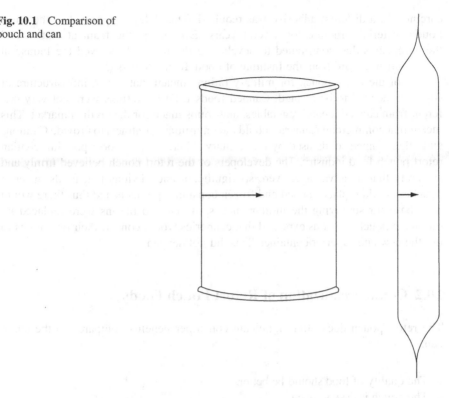

the seams of metal cans or the seal of glass jars. Such post-processing contamination is a major source of failure in canned foods. The usual precaution is to chlorinate the cooling water to reduce the chance of microbes entering through such leaks.

The challenge in developing the retort pouch was to fabricate the container from a material that was flexible, strong enough to resist high temperatures and high internal pressures during sterilization, and still provide a good barrier to oxygen and moisture. No single material was adequate, but a laminate of three materials was found to satisfy the requirements. The interior layer, in contact with the food, is polyethylene or polypropylene, thermoplastics that can be heat sealed by fusing with heat. These polymers are not especially good barrier films. A thin layer of aluminum foil provides good protection against oxygen and moisture transmission. Finally, there is a thin layer of polyester, which protects the aluminum foil and provides a surface on which to print graphics and written information. It is important to remember that food packages not only contain and protect their contents, but also they convey information and contribute to marketing.

Laminates are fabricated by using adhesives to hold the layers together. The first adhesive used to make the early retort pouches was a polyurethane compound that contained small amounts of unreacted monomer. Under severe conditions of high temperature, it was found that the unreacted monomer could migrate through the inner layer and, in theory, enter the food. Since the monomer was a potential carcinogen, a different adhesive was required. This delayed approval of the retort pouch material structure for several years. Eventually, the team at Natick and the companies that contributed to developing the material received the Industrial Achievement Award from the Institute of Food Technologists (IFT).

Use of the retort pouch for military rations meant that a new infrastructure of suppliers needed to be created. Canned foods used for rations were not very different from canned meats, vegetables, and soups made for the civilian market. This meant that commercial canneries could bid on military contracts to provide C rations and other canned foods used by the military. There was no corresponding civilian retort pouch food industry. The developers of the retort pouch believed firmly and sincerely that its advantages were so significant and obvious that foods in retort pouches would replace canned and frozen consumer products and thus there would be a base for supplying the military needs. As canned rations were replaced by rations in pouches, it was expected that canneries would convert their operations to use the new and superior container. This did not happen.

10.2 Commercialization of Retort Pouch Foods

The retort pouch does offer significant consumer benefits compared to the metal can:

- The quality of food should be better.
- The pouch is easy to open.
- There should be less energy consumed in its processing.
- The contents can be heated by immersing the pouch in boiling water, avoiding the soiling of a pan.
- The used pouch is easy to dispose of.
- A large number of unfilled containers can be shipped in a very small volume.
- It was thought that plastic costs might increase more slowly than those of metals.

Several manufacturers of processed foods, such as Green Giant, Hormel, Swift, and Kraft, investigated the new container and developed lines of foods to use it, usually based on existing products, such as chili and beef stew. Some of these same companies either were already producers of military rations or also evaluated the potential of that market. The US government encouraged entrepreneurs, especially minority and women owned enterprises, to establish ventures to produce retort pouch foods for rations. Few of these had any substantial food processing experience.

It should be noted that retort pouch foods make up only a part of a complete ration kit. The official name of the military ration is Meal Ready to Eat (MRE) and it consists of a fairly large polymer bag into which are placed an entrée in a retort pouch, a pouch of powdered beverage such as instant coffee, pouches of sugar and coffee whitener, paper napkins, crackers, dessert, matches, candy, gum, plastic utensils, and other dried or shelf-stable meal components, such as fruit, vegetables, and starch. There can be 12–15 items. Ration assembly is a separate operation sometimes performed by the same firms that make retort pouch foods and sometimes by others.

ITT Continental Baking Company (CBC) at the time in question, the late 1970s, was a diversified food manufacturer within a very large and complex conglomerate. The company was best known for its Wonder Bread and Hostess snack cakes (cupcakes and Twinkies, among others), but also owned Morton Frozen Foods, Gwaltney Meats, C & C Cola, Pearson Candy, and Paniplus food ingredients. CBC decided to create a new business unit, Continental Kitchens, to make shelf stable entrees in retort pouches competitive with high-quality frozen meals. The initial plan was to focus on the civilian market because of an assumption that military contracting was unprofitable and carried onerous bureaucratic complications.

The delay in FDA approval of the packaging film was advantageous as it permitted the relatively inexperienced (in this area of the food industry) ITT CBC to catch up with those firms that had started investigating retort pouch foods earlier. ITT CBC had some relevant product development expertise in its Morton division and had a flexible pilot plant. They also had some experience with flexible packaging for baked goods.

Commercialization of retort pouch foods at Continental Kitchens involved nearly every step one can imagine. Engineers and food scientists learned new skills and people with relevant experience were selectively added to the staff. New equipment including two types of retorts, a small filling and sealing machine and preparation equipment were procured. Preformed pouches from several suppliers were tested. At the time, there were three suppliers of packaging film, which could be provided in roll stock or as premade pouches, where three of the four seals were made by the manufacturer and the fourth by the food company, after filling. Using roll stock requires that all the seals be made by the food company on a more complex machine, but there is a significant cost premium for using premade pouches. The business decision was made to use roll stock and a manufacturer of form-fill-seal equipment was contracted to build a machine that would be acceptable to the US Department of Agriculture (USDA), since the products contained meat.

There were several suppliers of retort pouch packaging equipment, most in Japan or Europe, where retort pouches, using the early adhesive, were in wide use. The equipment supplier selected by ITT CBC was American and was accustomed to building form-fill-seal machines for packaging dry foods in flexible film packages. They had not previously built a machine adapted for wet cleaning – wash down – as is required when processing meat-containing foods. Equipment used in processing meat must meet the requirements of the USDA, whose requirements typically include materials of construction that resist corrosion and design that facilitates inspection. In practice, this means that most meat processing and packaging equipment is constructed of stainless steel, whereas most dry food packaging equipment is built with painted carbon steel frames. The form-fill-seal equipment that was designed for retort pouch foods was made entirely of stainless steel and was much more expensive than the other units made by the equipment company.

A major issue in evaluating systems for retort pouch foods was the speed of packaging. Cans are routinely filled at rates of 600/min and higher, making the process quite productive, since one or two people are all that are required to manage such an operation. Retort pouch form-fill-seal machines were capable of rates of 60/min. As the prototype machine was being built, work commenced on development of a higher speed machine targeted to achieve 250/min.

ITT CBC conducted several small scale market tests using products made in the pilot plant. There were six varieties: beef stew, Chinese pepper steak, chicken ala king, veal scaloppini, beef burgundy, beef stroganoff, and chicken cacciatore. Prices (in 1977) were generally more than $3 for an 8-oz serving. The novelty of the products, and heavy purchases by curious competitors, led quickly to demand exceeding the ability of the pilot plant to keep up the supply. One valuable lesson from the experience is that most pilot plants and most research personnel are poorly suited to function as manufacturing facilities.

The small market tests were terminated and plans made for a larger test, using a co-manufacturer. It was decided to choose a co-manufacturer who already had retorts, as these represented a significant capital cost. A vegetable canner in Pennsylvania was chosen, a contract negotiated, and equipment for food preparation selected. The prototype form-fill-seal machine was installed along with a carton packing machine and an inspection line.

In the early days of retort pouches, a major concern was seal integrity. The concern was that entrapped food particles or wrinkles could form leaks through which contaminated cooling water could enter. The seals also weaken when they are heated to retort temperatures and internal pressure could cause seal failure. To counter that effect, retorts use air pressure during cooling. The retorts available at the co-manufacturer were vertical, steam agitated vessels, a relatively old-fashioned but reliable design, as compared with horizontal retorts that are more common in new canneries. This mattered, it turns out, because the older retorts are more difficult to control and usually result in significant overprocessing of foods sterilized in them. This negates one of the expected advantages of the retort pouch.

About this time in the story, I was appointed director of process development and engineering and made responsible for product and process development of Continental Kitchens. I inherited all the previous decisions: the product line, the name, the graphics (deliberately intended to resemble the packaging of a successful frozen food line), the form-fill-seal machine, the staff, and the co-manufacturer. I recruited some food scientists and engineers and together we embarked on quite an educational excursion.

10.3 Issues and Lessons

Thermal process determination. In thermally sterilized foods, it is necessary to establish the rate of heating for each formula in each container and in each specific retort. The temperature distribution in the retort must be established with a realistic load of filled containers. Usually the containers are filled with a bentonite slurry and are used over and over for this purpose. Bentonite is a finely ground clay and the slurry simulates the density and viscosity of a thick fluid food.

The rate of heating the food is determined by placing thermocouples in the slowest heating position, usually the center or, in a can, just below the geometric center. This is easy in a rigid container, but how is it done in a flexible pouch? We devised a technique of sealing a piece of pouch material to the two sides of a pouch in the middle and attaching the thermocouple in the center, so that if the pouch expanded, the thermocouple was always in the center of the food. In practice, pouches are placed between perforated sheets of plastic to maintain their shape.

Another issue in thermal process determination is the practice of testing under extreme conditions. For our products, this meant overfilling the contents, bundling meat pieces (to simulate clumping), and using excess starch to increase viscosity, which reduces heating rate. We determined the impact of each of these conditions and argued, successfully, to do our testing with only the most severe situation, not the combination of all of them. Our contention was that the odds of all occurring at the same time were very slim. Our objective was to deliver a thermal process with a lethality value, F_o, of 7.0. For comparison, most meat-containing foods are thermally processed to an F_o of 6.0. The USDA dictated the higher value because of uncertainty about the retort pouch in those early days. When we measured actual heating rates of correctly formulated products in the co-manufacturer, we determined lethality values of 14–20, much higher than necessary. This meant we were not achieving the efficiency of thermal processing that was thought to be one of the big advantages of retort pouches.

Seal inspection. Pouch seals were inspected visually for defects. We determined in the pilot plant how fast pouches could be presented to inspectors with a high probability of accurate identification of defects. We had selected pouches with naturally occurring defects and could present streams of product with known defect rates. We determined that people could inspect at a rate that was about 50% of our production rate, so we needed at least two inspectors. However, we also found that visual inspection, while achieving a high probability of accuracy, did not achieve 100%, so we used three levels of inspection. These practices were adopted by the retort pouch industry and may still be used, since there still is no reliable instrumental method for seal inspection.

We also determined that some seals with visual defects were perfectly sound while some real defects – inadequately heated seals, for instance – were not visually obvious. The latter condition could only be prevented by ensuring that seal bars were always at the correct temperature and that sealing jaws exerted the correct pressure. Regular calibration of temperature indicators and of seal bar alignment became critical. Seals were studied by cycling pouches with various apparent defects under vacuum and pressure in an inoculated broth, so that if a leak occurred, the pouch contents would be contaminated and would support microbial growth, as shown by swelling due to gas being produced. Photographs of acceptable and unacceptable seals were provided to the USDA and became industry standards that may still be in use.

Packaging machine commissioning. Most packaging equipment is complex with many individual operations occurring very quickly and in precise sequence. In the case of our retort pouch form-fill-seal machine, the following are some of the steps. There were 14 stations arranged horizontally on a common chassis. An electric motor drove a common shaft through a proprietary Geneva drive that provided intermittent motion. This means that the pouches started and stopped at each step, as distinguished from continuous motion.

- The continuous roll of film passes over a smooth plow and is folded in half.
- The side seals are formed by a hot bar. (Hot bars may be heated continuously or be impulse heaters, in which current is applied just during the sealing event.)
- The bottom seal is formed by a hot bar.
- The seals are cooled by a water-cooled bar.
- The side seal is cut in half vertically creating a pouch, which is grasped near the top of the pouch by clips that travel on a continuous chain.
- Registration of the graphics is checked before the cutoff by an optical sensor detecting a black bar printed on the pouch. Registration means that the graphics are properly centered on the package. Some inexpensive packaging avoids registration issues by printing graphics smaller than the package size, so a portion of the graphics is always present, even if the packaging material stretches.
- The pouch is opened by moving the clips toward each other and blowing compressed air at the top.
- Meat (or another solid) is dropped in.
- Sauce is deposited using a volumetric dispensing valve, which has a vacuum suck-back to prevent dripping onto the seal area.
- The clips draw away from each other to form a flat top seal area, which is then sealed with a hot bar.
- The top seal is cooled.
- The clips open and the pouch drops to a belt conveyor.
- The pouch passes over a check weigher, a scale that records the weight of each pouch and rejects, with a puff of air, any that are under or over target weight.
- The pouches are collected and placed by hand in layers in a retort basket. The layers are separated with perforated plastic panels.

Each of the operations on the packaging machine is controlled by cams driven by the rotating shaft. Timing is obviously critical. We used high-speed video recorders to slow down the action and determine where slight adjustments needed to be made. The operations of opening and closing the pouch could create wrinkles in the seal. Poor coordination between the filling valve and the opening of the pouch could result in loss of a pouch.

We determined that the stainless steel chain that was installed as part of the design for wash-down conditions was inclined to stretch slightly over time and then caused malfunctions of the machine. The stainless steel chain was replaced by a more conventional carbon steel chain, which was less prone to stretch.

We had started the commissioning process conventionally, by trying to run slowly, at 30 pouches/min, with the intention of progressing step by step until we reached the target of 60/min. After the first few steps, we realized that at each step we were fixing different problems. We resolved then to jump immediately to the target rate, where we then remained for the year or so we operated. The lesson I learned is that many packaging machines have relatively little flexibility or turndown ratio – they are meant to run at their design rate and no other.

Marketing lessons. There was price resistance at the retail level. Even now, $3/package can seem expensive for an unknown product. Even at that price, costs were still higher because the meat ingredients were expensive and the volume of production was very low. Later, we developed another line of products that were designed to sell for around $1, a much more palatable price.

It was never clear where the product belonged in the grocery store. We did not quite appreciate the fact that the products were a new category. Superficially, they were somewhat similar to canned beef stew, canned pasta and sauce, or canned chili, but those products are mature and do not carry the quality image we wanted to convey. Some of our products had no shelf-stable counterpart. We did not want them near the frozen food section because we worried that shoppers might think they should be frozen, even though we viewed frozen entrees as the major competition. We had claimed quality equivalent to frozen, but with greater convenience. Finally, we discovered the products sold best if located near the meal helper section. Meal helper products are considered newer and more interesting and, while our entrees were complete, they seemed to fit well in that environment.

As an experiment, we made some products that contained just the sauce, having observed that our meats not only were expensive, but also in the thermal process the chicken actually got tough and the beef was not very good. We reasoned that we could sell a nice Stroganoff sauce and people could add their own hamburger or steak, as they chose. ITT CBC never pursued that concept, but other food companies did later.

Having recognized the price resistance to our entrees, we developed a new line addressing a new meal – lunch. The line was called "Deli On Your Shelf" and included five products: Sloppy Joe, mayonnaise potato salad, German potato salad, meatballs in sauce, and beef in barbecue sauce. They retailed for about $1/package and had a bright yellow carton. They were tested briefly in Atlanta and were, apparently, quite popular with college students. While Deli On Your Shelf never was exposed to a national audience, it still represents a classic case of identifying a unique application of a new technology in response to consumer information.

Military rations. After having avoided the military market, we investigated further and determined that we could use our excess capacity profitably. The three test markets required a few hundred thousand pouches per year, but the equipment and facility were capable of producing more than ten million in two shifts. The ration entrees had different recipes and package weights than our entrees did, but the pouch width was the same, so the machine needed little adjustment. We developed prototypes of several products according to the recipes provided by the military, got approval, and negotiated prices and quantities. Sales to the US government are typically based on allowable cost recovery and a fee for overhead and profit. We offered to receive a fixed fee, as distinguished from a percentage fee, so that we did not care which products we made. Otherwise it would have mattered whether we made baked beans or ground beef in gravy. We received orders for about nine million pouches and made enough money that we largely recovered our investment in the project.

10.4 What Happened?

ITT CBC decided not to invest the $25 million (by our estimate) that a national scale plant would have required and instead invest in new bakeries. Continental Kitchens was sold to another food company who discontinued the retail products and continued making pouched foods for military rations for a while. Then peace broke out, the demand for military ration components declined, and the operation was closed.

Retort pouch foods have never achieved the success in the United States that was expected. There have been periodic attempts to apply the technology to food products, including baby foods, pet foods, weight loss diet items, and meal kits containing sauces and starches, such as cooked rice. Three or so wars have demonstrated that retort pouch foods are like machine gun bullets, in the sense that only the military really needs them. The few companies that supply the military needs have few or no other markets. In times of high demand, the film for retort pouches can be in short supply because most is made in other countries. At times, the military have purchased commercial products, such as entrees in microwaveable plastic trays, because they could not get enough food in retort pouches.

Most retort pouches today are gusseted so they can stand on their own and do not need an outside carton. The gusset is an extra fold at the bottom that forms a stable bottom. The design uses a little more film than a plain "pillow" but does not need a carton and provides a better shelf presentation. Some beverages are sold in gusseted pouches.

One of the more popular applications of retort pouches has been for tuna and chicken. These products have canned counterparts and the products in pouches are noticeably superior in the opinion of many people. That observation is probably key to understanding why the retort pouch has not been a more successful food package in the US *Foods in retort pouches are superior in quality to the same food packaged in a metal can*. This is true for their initial application – military rations – and for thermally preserved tuna and chicken. Foods in retort pouches are not equal in quality to the same products sold frozen.

Finally, retort pouches with aluminum barrier layers are not microwaveable. Plastic trays and tubs that are microwaveable are used for thermally preserved foods that might have been candidates for packaging in retort pouches. There are laminated films that do not use aluminum barriers, and so can be heated in a microwave oven, but these are relatively expensive and not in very wide use.

In summary, the retort pouch is a significant advance in food processing and packaging, but it has not been the commercial success it was expected to be. Nonetheless, its commercialization embodies nearly every challenge and lesson that a new food product can have.

10.5 Exercises

1. How might filling and sealing flexible containers be improved? Options include vertical or horizontal orientation; rotary or straight line; multiple lanes; and "bandolier," in which pouches are not cut apart until after they are filled. All such options have been applied, though not necessarily for retort pouches. Discuss the apparent advantages and disadvantages.
2. Discuss whether agitation of pouches during sterilization will improve heating rate.
3. Ultrasonic sealing of pouches has been demonstrated. Discuss the advantages and disadvantages.
4. Seal inspection and non-destructive testing remain challenges. Discuss possible approaches.

10.6 Lessons

1. Just because a technology or product has many qualities does not mean it will be appreciated or successful.
2. Few thermally processed foods have sensory quality equivalent to the same food frozen.
3. Introducing a new brand, a new package, or a new food product category is each challenging alone; doing all three at once, as ITT CBC attempted with ready-to-eat entrees in retort pouches, is nearly impossible.
4. Depending on an item with no real civilian market, as the military does with rations and ammunition, requires the equivalent of dedicated arsenals or food plants that must be supported when demand is low. (Surplus rations are often used as emergency supplies in cases of natural disasters, such as floods and earthquakes.)

Chapter 11
Ice Cream

Ice cream is not only a popular dessert but also one of the more complex foods we eat. There are four physical phases in ice cream: solid ice crystals, fat globules, droplets of super-cooled aqueous solution, and bubbles of air. In addition, there often are suspended inclusions such as nuts, fruit, pieces of cookie dough, and streaks of syrups, such as caramel or chocolate. Ice cream is normally made from cream, sugar, flavors, and emulsifiers, but it can also be made from butter fat and non-fat dried milk in places where fresh cream is not easily available. We think of ice cream as a hot weather treat, but some of the highest per capita consumption occurs in cold countries such as Scandinavia. Normally, ice cream plants are located near a good supply of milk and the product shipped to more populous markets.

11.1 Ice Cream Manufacture

The ice cream manufacturing process is shown in Fig. 11.1. Cream is separated from raw milk by centrifugation and delivered by tank truck to the ice cream plant, where it is stored in silos. Normal practice is to clean storage silos every 72 h, so enough silos must be provided and their use must be carefully scheduled to permit completely emptying each one as needed.

Cream (or a re-constituted substitute made from butter fat, non-fat dry milk (NFDM), and water) is combined with sugar, emulsifiers, some flavors, and additional water, if needed to dissolve soluble components, in an agitated batch mix tank. The mix is pasteurized, normally in a high-temperature, short-time (HTST) continuous thermal process with regeneration, as discussed previously. The mix is more viscous than fluid milk, and so the clearance between the plates should be wider than it is in plate heat exchangers designed for milk or juices. The wider clearance reduces the pressure drop through the exchanger, also reducing the stress exerted on the gaskets sealing the plates. The wider clearance also means that heat transfer coefficients are less than in other plate heat exchangers, so more area is normally required for the same heat transfer duty.

As is common in other pasteurization of dairy products, a homogenizer is normally installed after the hold tube and before the regeneration section so as to treat

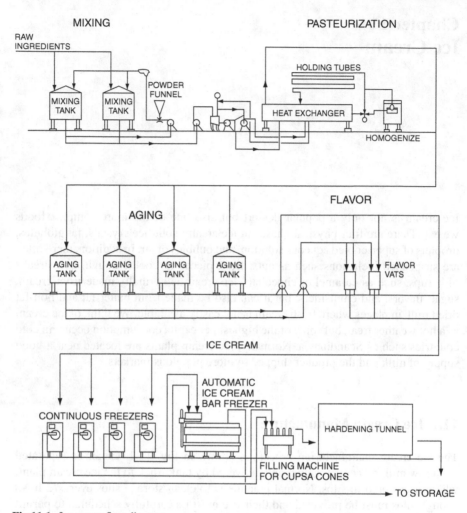

Fig 11.1 Ice cream flow diagram

the hot fluid and also provide repressurization to maintain the pressure of the treated fluid higher than that of the raw stream.

Ice cream can also be pasteurized by heating in batch tanks for longer periods of time, up to 30 min, as compared to 30 s in HTST pasteurization. The longer, slow heating can cause detectable flavor changes in the mix, specifically caramelization of sugars and a slight darkening of color due to the Maillard browning reaction between sugars and amino acids from the milk protein. For most ice creams, the flavor and color changes induced by batch pasteurization are undesirable, but in other cases, the results of batch pasteurization have become identified with specific products and consumer preferences require that the somewhat old-fashioned process be continued in use, or simulated by using severe conditions in an HTST pasteurizer – higher temperatures and longer hold times.

Fig. 11.2 Ammonia refrigeration system

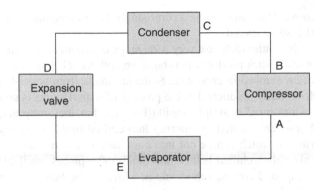

After pasteurization, the mix is sent to continuous freezers, which are wiped surface heat exchangers normally cooled by direct expansion of ammonia or by very cold glycol solution, which is chilled by ammonia. Ammonia is normally used for refrigeration because it is an efficient refrigerant for large capacity central refrigeration systems. In an ice cream plant, there are at least two levels of refrigeration required and often three – one for the continuous freezers, one for the hardening system, and one for the storage freezer. See Fig. 11.2.

Ice cream mix leaving the continuous freezers is a thick fluid, essentially the same as "soft serve" ice cream. If there are to be inclusions, such as nuts, fruits, or syrups, they are added at this point. It is important to note that the inclusions are not normally heat treated and so pose a potential source of contamination. This is not usually considered a hazard because ice cream is preserved by freezing, but if ice cream is abused by allowing it to thaw and then is re-frozen, it is conceivable that pathogens introduced by the inclusions could grow and cause illness. To prevent this admittedly remote possibility, some inclusions are sterilized when that is feasible. For example, the flour used to make raw cookie dough can be sterilized by long slow heating – 140°F for 24 h – without impacting its functionality. Some fruits and syrups can be packed aseptically. Some frozen fruits, such as strawberries and cherries may be blanched before freezing and have sugar added to reduce their water activity, all of which can reduce any microbial load. Nonetheless, it is important to realize that ice cream can pose a health hazard if it is mistreated.

Fortunately, ice cream that has been abused often will show visual and textural evidence. If ice cream is thawed and refrozen, ice crystals can grow and become visible. The solubility relationships among the components of ice cream are delicate – the aqueous phase is often close to saturation and so proteins and sugars can precipitate, causing a condition known as sandiness and sensed as a gritty texture in the mouth.

After freezing and after inclusions are added, the soft ice cream is pumped to filling machines or used to make novelties. Ice cream packages can range from single serve cups of a few ounces to five gallon tubs used in dipping stores and food service. Often the packages are made of coated paper. Filling machines for packages usually have drips and small spills and often use running water to wash these spills away. This can become a surprisingly large consumer of water and a contributor to the waste stream.

Novelties include a very wide range of product forms including bars, cones, sandwiches, stick products, bon-bons, and others. They typically are single-serve pieces, often enrobed in chocolate. Some are made from water-based formulas rather than from dairy products, but the process of manufacture is similar. The base is usually "white mix," a simple vanilla formula, but they may also be chocolate flavored. Forms are created by pouring into chilled molds or by extrusion of a continuous ribbon, which can be cut into bars. Bon-bons are pieces created by dropping soft mix onto a chilled belt or into molds. The pieces are then enrobed in chocolate or compound coating (chocolate in which cocoa butter is replaced by less expensive fat). Chocolate or compound coating for use on ice cream must be formulated so as not be too brittle at low temperatures, which can be a challenge.

After package filling or fabrication of novelties (which are usually packaged in cartons), ice cream is hardened by cooling to about − 40°F (−40°C). Hardening can be continuous, in tunnels or spiral freezers, or batch, holding in blast freezers. Blast freezers are insulated rooms in which cold air is circulated by fans to improve external heat transfer. Tunnels or spiral freezers use enclosed belts supplied with cold air. A spiral is specially constructed to reduce floor space by driving a woven belt around a central shaft. Normally spirals are operated in pairs so that entrance and exit are at the same level.

11.2 A Digression into Heat Transfer

Ice cream packages are normally hardened individually, but some other instances of heat transfer are applied to larger units, which can be a disadvantage. Some examples include cooling of refrigerated foods, incubation of yogurt in cups, leavening of refrigerated biscuit dough in tubes, and freezing of food service muffins.

11.2.1 Cooling Refrigerated Foods

Foods maintained at reduced temperatures may have higher quality than their shelf-stable counterparts because they typically undergo a less severe thermal process. They may receive a premium price because of the perceived higher quality and the higher cost of a refrigerated distribution chain. One example is refrigerated salad dressing, which may have a higher pH (be less acidic) than shelf-stable salad dressing. In at least one case, refrigerated salad dressing is mixed, emulsified, filled into glass or plastic jars, placed into cases, the cases placed on pallets (fairly tightly packed together), and then the pallets placed in a storage freezer. It takes multiple days before all the jars reach the desired temperature of about 40°F. Normally, when the objective is to cool a food, it is desired to do so as fast as possible because harmful reactions can occur while the food is warm.

11.2.2 Fermentation of Yogurt in Cups

Yogurt is made by fermentation of sterilized milk with selected microorganisms. It can be done in batch tanks or in individual cups. The product fermented in cups can have a different texture than that fermented in tanks and is vulnerable to contamination during filling. After filling with inoculated milk in a clean environment, the cups are sealed, placed into cartons, the cartons placed on pallets, and the pallets placed in warm rooms for incubation. The pallets are held for several days until the proper texture is achieved and then cooled for distribution.

11.2.3 Leavening of Refrigerated Biscuit Dough in Tubes

Refrigerated biscuit dough is a great convenience product that is made by mixing a dough of flour, water, and baking powder and then by continuously rolling and folding the dough into a ribbon or sheet. When the sheet is the proper thickness, after usually three levels of reduction, individual disks are stamped with dies and stacked into paperboard tubes. The tubes are capped, placed in cartons, the cartons placed on pallets, and, sometimes, the pallets placed in warm rooms to partially activate the leavening agents. (Commonly, the dough is adequately leavened by its time at room temperature during sheeting.) This causes the dough to expand within the tubes and prepare the disks to become biscuits when the tube is opened and the dough placed on metal sheets and baked. The quality of the biscuits – the uniformity of their response to baking – depends on each receiving the identical degree of leavening.

11.2.4 Freezing of Food Service Muffins

Muffins are prepared by mixing a batter of flour, sugar, oil or shortening, leavening agent and water, pouring the batter into paper-lined molds, baking in the molds, cooling, removing the muffins by hand, wrapping each individually, placing the muffins in cartons, placing the cartons on pallets, and putting the pallets in a storage freezer. (Batter is thinner than dough and can be pumped and deposited through valves.) It can take more than a week for the muffins to become frozen. Most baked cereal products become stale, indicated by becoming hard, when held at reduced temperature for even a short time.

These examples have in common the fact that food products are subjected to heat transfer – heating or cooling – in relatively large geometries – cases on pallets – and probably suffer in quality as a result. The rate of heat transfer is governed by conduction in each case because the cases are packed tightly together on the pallet.

In conduction heat transfer, the time to achieve a given target temperature is directly proportional to the square of the characteristic distance, in this case, the distance to the center of the pallet. Pallets are usually 40×44 in., so the average distance to the center is about 21 in. In contrast, a typical case might be 6 × 18 × 24 in., so the distance to the center could be 3 in. Individual containers might be about 3 in. in diameter, so the distance to the center is about 1.5 in. This means that the relative times to achieve the same target temperature, starting from the same initial condition, is 441 for pallets, 9 for cases, and 2.25 for individual units. The specific geometries and physical properties will determine the actual times, but the point is that treating individual units or cases, as is done in ice cream hardening, is far more efficient than treating full pallets. Not only is quality improved, but also the time in which valuable product is held is greatly reduced. In some cases where this concept has been applied, the estimated benefits were substantial, at relatively little cost.

Why would anyone treat full pallets instead of individual units or cases? First, it is more convenient to handle pallets than individual units. Material handling is a major cost in food plants. Second, the appropriate heat transfer device may not be available. Rooms for cooling, freezing, or warming are simple to build and operate, whereas a continuous tunnel is more complex and occupies valuable floor space. A spiral belt machine is more compact than a linear tunnel, but is still a significant investment. Individual units could be placed in a room, but the spatial loading would be less than if they were packed on pallets, so more rooms would be required for the same capacity. Even though the time is greatly reduced, there is an increased space requirement. Thus, there are rational reasons why heat transfer is performed on pallet loads rather than on individual units or cases. However, wherever possible, it is probably an improvement to perform heat transfer in the more efficient fashion.

11.3 Sanitation and Cleaning

The concept of cleaning in place (CIP) was developed for dairy processing and later applied to other food manufacturing (Seiberling 1997). Before automated CIP was applied, it was customary to disassemble a dairy processing plant and clean each piece by hand. Dairy products provide a hospitable environment for microbes to grow and are prone to leave films on hot surfaces, such as heat exchangers. These films are composed of denatured proteins and can be difficult to remove. Cleaning agents are usually caustic and acid. A typical cleaning sequence involves rinsing with water, circulating hot cleaning solution (normally 1.5% caustic), rinsing, circulating mild acid (usually nitric or phosphoric, since hydrochloric acid can cause corrosion in stainless steel), rinsing again, and finally circulating a sanitizing agent, often an iodophore, which is left in the lines overnight and displaced upon start-up with water.

Cleaning is usually the largest use of water in an ice cream plant and also the largest generator of liquid waste. Water conservation should always be a concern in a food plant, but in some arid parts of the world, it is a necessity, not an option. How can water be conserved in making ice cream? As previously mentioned, some filling equipment may use a continuous stream of water to wash away spills. This really is not necessary, so selecting equipment that uses less water is a good first step in reducing water usage.

Recognizing that cleaning is a major user, it is constructive to consider each step with a view toward conservation. The first rinse in the usual sequence also generates the strongest waste because this rinse removes the residual film of cream from tanks and equipment. One approach is to recover the rinse stream and to use it as ingredient water. Many ice cream formulas have added water to dissolve sugar and other ingredients. If the rinse water is treated carefully, it can be used instead of fresh water. The extra solids in the rinse water are better used in the mix rather than in the waste stream. Normally, no credit is taken for the contained solids in formulation because the concentration varies, but it is better to use them productively than to throw them away.

Finally, in most food plants, it is common to have many hoses that are used to clean spills and wash the exterior of equipment. (The interior of equipment and of pipes is cleaned automatically by spray balls or by circulating streams at high velocity.) Sometimes these hoses run all the time because they do not have valves on the end but only at the source of water. A good practice is to install "dead man" valves, which only open when attended, on each hose. (They are called "dead man" because they will not open unless they are actively operated – a dead man cannot do this.) It is also useful to examine critically where hoses are installed and to challenge how many there are.

Operators want hoses convenient to where they work, but the more hoses there are, the higher is likely to be the water usage. More hoses require a higher investment in utility piping. Hoses that are convenient are more likely to be used for purposes such as chasing spills to a nearby drain. A better practice is to encourage "dry" cleaning of spills, with brushes and pans or shovels. Such practices require cultural changes in workers, but can significantly reduce water consumption.

11.4 Examples

1. A super-premium ice cream was traditionally produced using batch pasteurization. (Super-premium ice creams typically have higher fat content and lower overrun than ordinary ice creams. The standard of identity for ice cream requires a minimum of 10% butter fat and a maximum of 100% overrun. Overrun measures the volume of air incorporated in the mix. One hundred percent overrun means that the density of the mix is about 50% of its density before air is added. Air is added by injection after the freezer.) Upon its acquisition by a large food company, consideration was given to replacing batch pasteurization by the more modern HTST. By some standards, the flavor was improved, but consumers may view any change in the flavor of an established product negatively. Engineers for the new owners decided to modify the HTST to simulate the effects of the batch process. As an exercise, determine the likely conditions of the batch process, the normal HTST process, and the modified conditions that would give a "cook" comparable to that of the batch process.

2. In a multi-product ice cream plant, there may be at least two mixing and pasteurization systems, a dozen mix tanks, and as many as twenty freezers. Some lines

may have two or three freezers, while others have one. Large packages consume more mix in a given time period than do smaller ones. In a desire for flexibility, one owner required that any mix tank be capable of feeding any freezer. In an automatically cleaned system, the amount of piping is about twice what might otherwise be required because the cleaning solution is recycled. To easily route any tank to any freezer, valves are arranged in groups and remotely operated. As an exercise, sketch how such a valve arrangement might look. How many valves are required?

3. Valve matrices can be located near the floor or elevated. Why choose one over the other? A location near the floor is convenient for maintenance. The type of valves involved requires periodic maintenance because they have replaceable seals, which can wear out. However, a location near the floor then requires vertical runs of pipe to avoid blocking aisles and passage ways. Vertical legs of pipe are difficult to clean. Locating valve clusters overhead reduces the complexity of the piping system at the cost of more difficult access. An interesting and attractive solution to this dilemma is to locate the mix tanks on a second floor, above the freezers. The valves can be easily accessible and drops of piping to the freezers are mostly vertical, but the flow is down and thus more easily cleaned. An additional benefit is that pumping distances are shortened, which reduces air incorporation, an important consideration in making super-premium ice cream. At least one plant for super-premium ice cream was constructed this way.

4. A very large, multi-product ice cream plant was designed for an arid area, using many of the water conservation techniques described earlier. During construction, another company acquired the plant's owner. The engineers for the new owner were not fully aware of the considerations involved in the original design and, among other changes, installed additional hose stations, at the request of the operators. They then complained to the designers that the utility lines were undersized. Upon inspection, it was realized that there were about twice the hose stations that were originally intended. The water conservation motivation was explained and the new owners agreed to reduce water use consistent with the original intention.

5. The normal practice for CIP of pipe systems in the United States is to require fluid velocity of at least 5 ft/s (about 1.5 m/s). The intention is to ensure that flow is turbulent. In Europe, it is recognized that turbulent flow is dependent upon both velocity and pipe diameter. As an exercise, calculate the velocities required for 1.5, 2, and 3 in. tubing, assuming constant Reynolds Number and compare the results with US practice. What Reynolds number should you assume? What are the economic consequences of varying velocity?

11.5 Lessons

1. Automated cleaning was developed originally for dairy plants and has been extended to other food operations. It is one of the great developments in food engineering and requires creative application of fluid flow.

2. Excessive flexibility can lead to excessive complexity and rarely is really required. In the example cited, a few of the possible tank to freezer connections were actually used and many were later removed.
3. Creative use of the third dimension, such as a second or third floor, can simplify flow systems and favorably impact quality.
4. Application of the principles of conduction heat transfer, by treating the smallest units possible, can reduce work in process and improve quality at relatively little cost.

Chapter 12
Sausages and Other Meat Products

There are many varieties of sausages made from beef, pork, poultry, and game. To understand this category of value-added meat processing, it is useful to consider meat processing in general. The meat industry is one of the largest segments of the food industry and also one of the most labor intensive and dangerous. While many animals are domesticated and used as food, the major focus in the United States is on beef, pork, and chickens. Sheep and goats are important in many cultures, and other animals are used elsewhere, including horses, dogs, snakes, and wild game.

Meat is often sold fresh, preserved by refrigeration; frozen; or further processed and preserved by several principles that may be applied, including cooking, reduced water activity, and chemical preservation. Meat products have been implicated in many recalls and outbreaks of foodborne disease, so sanitary design and operation of meat plants are especially challenging.

12.1 Basics of Meat Processing

Animals intended for human consumption must be healthy and alive at the time of slaughter. This means that animals that die of other causes, such as disease, old age, or some accident may not be used for food. The carcasses of slaughtered animals intended for interstate commerce are inspected by veterinarians or specially trained inspectors employed by the US Department of Agriculture Food Safety and Inspection Service (USDA FSIS). This visual inspection is directed at signs of disease, such as tumors, lesions, or malformed organs. Visual inspection alone cannot detect microbial contamination. Animals slaughtered at smaller plants intended for local consumption are generally governed by state public health or agriculture regulation and inspection. Examples include facilities used for dressing deer and other wild game.

Federal regulations dictate that food animals be killed humanely, meaning that they should not suffer unduly. This usually means that the animals are stunned before they are killed by having their throats slit. Cattle and hogs are stunned by electric shock or a blow to the head. Kosher and Halal ritual slaughter – for observant

J.P. Clark, *Case Studies in Food Engineering*, Food Engineering Series,
DOI 10.1007/978-1-4419-0420-1_12, © Springer Science+Business Media, LLC 2009

Jews and Muslims, respectively – do not allow stunning before an animal is slaughtered. Rather, the animals are rapidly inverted so that blood rushing to their heads renders them unconscious before their throats are slit by specially trained clergy – rabbis or imams – who offer a short prayer first. Federal regulations specifically define religious slaughter practices as humane. Typically, if Kosher or Halal meats are produced in a plant that also produces conventional meat, there are separate facilities and staff, which add to the costs.

After slaughter, the animals are hoisted upside down with a chain around one leg and hung from a hook traveling on an overhead conveyor chain. The conveyor spacing and height above the floor is determined by the species involved, which have different weights and lengths. Cattle generally weigh about 1,000 lb at slaughter and hogs are usually about 200 lb. Yields of edible meat are about 50%.

The economics of meat packing depend upon recovering value from as much of each animal as possible, so after the animal's throat is slit, it is common to collect the blood by allowing it to drain into a special system of pipes and tanks. Blood deteriorates quickly, so it must be preserved by refrigeration if it is intended for human consumption. Inedible blood is commonly dried and used as an ingredient in glue.

Meat packing is a disassembly process, starting with the live animal and ultimately resulting in many different products such as roasts, steaks, ground meat, sausages, various organs, and inedible by-products. As practiced in large meat processing plants, meat packing consists of many specialized tasks performed by hand labor with mechanical assistance. This can be contrasted with classic butchering, performed on one animal at a time by one or two skilled individuals. For many years, the industry practiced an intermediate division of labor in which animals were slaughtered in central facilities, often at transportation hubs such as Chicago, Kansas City, and St. Louis. Whole carcasses, halves, or quarters were shipped to population centers where butchers in local groceries or shops divided the large pieces into those used by customers.

The revolution in meat packing occurred when the functions of skilled butchers were broken down into individual tasks, which relatively unskilled labor could be trained to perform. Instead of shipping large pieces of meat, along with the less useful parts such as bones and fat, the fabrication of consumer level pieces was moved closer to where the animals are raised, such as Colorado, Nebraska, and Iowa. This move has many consequences. Skilled, and well-paid, city butchers lost their jobs. In their place, rural workers were hired at lower wages. Today, many of these jobs are held by immigrants from Mexico and Southeast Asia. New technologies have been applied, taking advantage of the large scale of operation. "Shelf-ready" cuts are packaged in central locations, reducing costs and increasing efficiencies in groceries. However, the large volumes also increase the risks of cross-contamination and have led to some very large and expensive recalls of meat, when pathogens have been found in centrally produced ground meat.

12.1.1 The Sequence of Disassembly

After the animal has been killed and bled, the head is removed, the animal is disemboweled and the carcass is usually split in half. The process depends on the heart continuing to pump to help remove all the blood. The head and entrails are put into traveling trays and kept with the rest of the carcass until they have been inspected, after which they go on to their own processing. The head meat is removed by hand in an area that often is the first assignment for new employees. Each person works on one head at a time, cutting off as much meat as possible. Some parts are specially valued, such as tongue, jowls, and cheeks, which may be smoked or pickled.

The entrails, after inspection, go down a chute or may be vacuum conveyed to a separate area, where they are cleaned and separated. The stomach and intestines become tripe or sausage casings, after extensive cleaning in equipment that resembles laundry washing machines.

The main carcass is split in half, hooves are removed, and organs such as the heart, liver, and kidneys are separated and sent off for packing or other use. The skin may be removed before the carcass is split or it may be left on until the carcass is cut into smaller pieces. If large pieces of skin are valued for making leather, the skin is removed early in special machines that grip it by an edge and pull it off. Removing the skin early can improve sanitation because most potential contamination is on the skin. Before removal of the skin, hog carcasses are scalded in hot water to soften hair, which is then beaten off with brushes. The carcass may also be singed with flames to help remove hair. Sometimes the hair is recovered and used for brushes.

Hog and cattle skins not intended for leather are used to make gelatin. The skins are covered with salt and refrigerated. Some organs from cattle and hogs are separated and used to recover hormones and other biochemicals.

Poultry slaughter and cutting follows a similar path, but is more automated because the animals are more uniform and lighter in weight. The birds are fitted into a carrier on a conveyor chain, stunned, and their heads removed. Feet are removed, the birds are scalded, and feathers removed by beating. The viscera are removed and conveyed separately. Carcasses are chilled quickly by immersion in a cold water bath. This is a source of contamination because the water is not completely refreshed. Chilling in cold air reduces the risk of contamination but is a bit more expensive because the birds lose some weight, while the water bath may actually add some weight.

Meat at or near body temperature is soft and difficult to cut, so normal practice is to chill carcasses of poultry, beef, and hogs before further processing. Cattle and hogs are typically hung overnight in refrigerated rooms. Poultry are chilled in cold water or air, as previously mentioned. Carcasses lose some weight in cold air, so the coolers may have water mists to raise the humidity and reduce yield loss.

The interior of meat is essentially sterile, but the surface may have a microbial population. Most such microbes are harmless, but some may be pathogens or can cause spoilage. The population that thrives at body temperature may be quite different from that which enjoys cooler temperatures. During chilling, the microbial ecology changes due to the changing temperature. In one case described later, advantage is taken of this fact to increase the shelf life of fresh sausage by hot deboning and very rapid packaging.

After chilling, the carcasses are cut into smaller and smaller pieces by hand labor working at tables with three central conveyor systems. One conveyor belt delivers primal cuts, another takes away smaller cuts, and a third removes wastes and trimmings. The workers are typically garbed in robes and wear head coverings, chain mail aprons, and chain mail gloves. They use ordinary knives or may use small saws and mechanized knives. The ordinary knives are sharpened on steels provided at each work station. There is also usually a sanitizing dip at each station with hot water that may have chlorine or another approved chemical added. The workers are close together, working fast and wielding sharp tools. Injuries are, sadly, frequent. In addition to cuts, repetitive motion injuries are common because the workers are performing the same tasks over and over. A common remedy is to rotate workers among several tasks, but this requires training and turnover is high in the industry, so a given worker may not be on the job long enough to learn more than one task.

Efforts have been made to mechanize some tasks, such as by using laser guided saws to make the first major cuts, which are often at joints. The small pieces of meat – steaks, chops, roasts, chicken parts, and others – are now packed in foamed plastic trays with film over wrap. These "case ready" packages are placed in larger cartons and may have a modified atmosphere of reduced oxygen provided to extend their shelf life. At the destination the large package is opened and the smaller packages then exposed to air. The film enclosing the small packages is permeable to oxygen. Red meats lose their red color in the absence of oxygen and regain the color when exposed to oxygen, but are also less likely to spoil because spoilage microbes usually require oxygen.

The trimmings and small pieces of meat not useful by themselves are sorted roughly by protein content and accumulated in bins. These become the ingredients for sausages and ground meat. Ground meat is sold in trays and chubs (plastic tubes) for consumer use and is fabricated into patties for hamburgers. Sausages are prepared and preserved in a variety of ways.

12.2 Meat Preservation

Meat products demonstrate many principles of food preservation including thermal processing (cooking), low water activity (drying, salt addition), high acid (fermentation), and chemical treatment (nitrites, smoke). Sausages and lunch meats often are prepared and preserved in similar ways but differ in the size of the product and how it is used. Sausages may be fresh, meaning they still require cooking, or ready to eat, meaning they have been cooked or otherwise made safe and edible. Sausages and lunch meats are made from all one meat – beef, pork, and chicken – or from mixtures of these. Exact formulas and flavors vary widely by culture and the manufacturer.

12.2.1 Forming

Sausages may be coarsely ground or finely ground. The latter often forms a smooth emulsion. Meat is usually ground while it is cold in order to preserve distinct particles. The meat may be frozen to start; ice can be added when the formula calls for added water; the grinder can be jacketed and cooled, though this is challenging; or carbon dioxide may be injected. Of these approaches, the use of carbon dioxide is most common because it leaves no residue, is most versatile and is effective. The act of grinding introduces energy to the meat and raises the temperature, causing smearing of fat unless the energy is removed.

A given formula will call for trimmings blended to achieve a target fat content. Fatty trimmings are usually less expensive than more lean (higher protein) trimmings, and fat is necessary to achieve the desired texture and flavor, but fat is more vulnerable to increasing temperature. Some distinctive products such as the Italian specialty mortadella rely on having visible and distinct fat particles in a smooth emulsified matrix, so these must be ground carefully and frozen fat pieces added near the end of the process.

Because chopped meat mixtures are viscous, mixing requires large amounts of energy, so the mixers and grinders are sturdy pieces of equipment with large motors. Salt and spices are added and must be well distributed. Some studies have shown that good mixing can require up to 20 min, but in practice, meat mixers rarely run for more than 2–3 min to avoid smearing and excessive size reduction. Emulsified products, such as frankfurters, are mixed for longer times and are more finely ground than more coarse products. The discrepancy between the time it would take for good mixing and the usual practice means that spices and salt may not be uniformly dispersed in a sausage mixture. If the mixture is stuffed into casings immediately, there could be considerable variation among pieces. It is common to hold ground mixtures in refrigerated storage overnight to allow some equilibration of the salt distribution and to extract some protein from the pieces of meat. The extracted protein can help bind the small pieces together in the finished sausage and improves texture.

Most sausages have the ground mixture of meat and spices stuffed into a casing or tubular material that forms the shape and may or may not be edible, contributing to the texture as well. Natural casings are made from the cleaned intestines of cattle, pigs, or sheep. These can vary in size and are edible. They are permeable to water and smoke, so permit some drying during cooking. Other products are formed in metal molds.

Edible synthetic casings are made from regenerated collagen. Collagen is the primary structural protein of animals and is extracted from bones and skin. It is dissolved in acidified water and then extruded in a thin walled tube into a caustic bath, precipitating the collagen. The casings may be dried, requiring rehydration for use, or kept refrigerated and moist. The great advantage of collagen casings is their uniform size, compared with the variability of natural casings. It is possible to co-extrude a collagen solution around a sausage mix into a caustic bath and thus generate an edible casing at the point of use.

Inedible casings are made from regenerated cellulose and other polymers. These are very uniform and relatively inexpensive. They can be perforated if necessary to permit penetration of smoke and loss of fat and water. Large diameter casings are used for some lunch meats, while others are made in loaf pans.

12.2.2 Cooking

Lunch meats in loaf pans are commonly cooked in hot water or steam, usually in batches, though there are continuous cookers that transport pans or casings through steam or hot water and then cooling water. Other small and large diameter sausages in casings are cooked in ovens or smoke houses which may be batch or continuous, with or without added smoke. Liquid smoke flavor can be added to formulas to eliminate the need to generate smoke in the cooking. Heating damp sawdust generates smoke. Temperature changes with time are determined experimentally and can be programmed into controls for the cooker. The time and temperature depend heavily on the diameter of the product, the type of casing, and the desired final texture. Most cooking protocols are designed to reach a minimum safe temperature at the center of the sausage, typically about 160°F. This is usually achieved by starting with lower temperatures and gradually raising the temperature until the target is reached. Smoking time and intensity is governed by the desired color. Both dry bulb and wet bulb temperatures are typically monitored, with the wet bulb temperature being an indicator of the relative humidity, which can affect weight loss due to drying. Some moisture loss is normal and desirable, but too much can affect the texture and increases cost due to yield loss.

Dry bulb temperature is measured with a thermometer or temperature indicator, such as a thermistor, while wet bulb temperature is measured with a similar instrument, which is enclosed in a wick saturated with water. Wet bulb temperature is lower that dry bulb temperature because the evaporation of water cools the instrument. The wet bulb temperature at a given dry bulb temperature correlates with relative humidity of the atmosphere.

Continuous cookers are most appropriate for large volume products, such as frankfurters, for which the conditions can be set once and then left alone for an entire shift. Batch cookers are more flexible and appropriate when many different products are being made.

After cooking, products are quickly cooled to avoid overcooking. Typically the products are sprayed with cold water or chilled brine. Products in inedible casings may be packaged and sold with the casing on or it may be stripped off. In that case, the products are briefly exposed to steam and the casing slit and peeled off. "Skinless franks" are the most common form of the popular frankfurter. These are typically loaded automatically into vacuum-formed plastic packages and sealed with a flat lid.

Ready-to-eat meat products are vulnerable to contamination after cooking and before packaging, especially by *Listeria monocytogenes*, a pathogen that thrives at low temperatures. Options to protect against this hazard include exposure to steam followed by vacuum to rapidly cool the product, spraying the product with organic acids, and heating the sealed packages in hot water.

Loaves cooked in pans and large diameter lunch meats cooked in batch ovens or smokehouses are sprayed with cold water or brine and, in the case of the loaves, unloaded onto conveyors. They may be packaged in large pieces for use in delis and sandwich shops or sliced into retail packages.

12.2.3 Fermentation and Drying

Traditional Italian salami and some other dry sausages are not cooked but rather are preserved by a combination of low water activity and natural acidification from fermentation. These products are often based on pork, which since it is not cooked, must be treated to remove the potential hazard of *Trichinosis*, a parasite of pigs that can cause disease in humans. *Trichinosis* is not very common, but precautions against it must still be taken. Because the parasite cannot survive cold temperatures, pork to be used in fermented sausages is stored in a freezer for several days before using.

The mixture is ground coarsely, mixed with salt and other flavors, and inoculated with a pure culture of lactic acid bacteria. The mixture is stuffed into a porous inedible casing and hung at a constant temperature and relative humidity. Traditionally, the curing and drying conditions were affected by the local weather, so artisanal sausages of this type were only made well in certain areas. By carefully controlling the temperature and humidity in closed rooms, using dehumidifiers and temperature controllers, the curing process can be accelerated from the common 30 days to less than a week. This has important economic consequences, since it is difficult to predict market requirements a month or more in advance.

The dehumidifier of choice uses a scrubber with a strong salt brine to remove excess moisture from the air. The brine is concentrated by heat and reused. Using a liquid scrubber has the added benefit of cleaning the recirculating air from extraneous bacteria that might contaminate the curing meat before it has developed its own protection.

What is happening during the curing of dry sausage? The lactic acid bacteria consume soluble sugars in the meat mixture and create organic acids. These lower the natural pH of the meat mixture below the level of 4.6 at which *Clostridium botulinum* can produce toxin. The acid also helps to denature the meat protein and change its texture. At the same time, moisture is lost through the porous casing, increasing the solids content of the mixture and lowering the water activity below the level at which pathogenic microbes can grow. The higher solids content also changes the texture of the mixture. Finally, the fermentation producing the organic acids also generates flavor changes in the meat. It is common for a white mold to develop on the outside of the hanging sausages, but it is not thought that the mold contributes much, if anything, to the flavor and texture. However, it is often considered a sign of the authentic product.

The diameter of the sausage has a significant influence on the time it takes to dry and cure. Very thin products, such as beef jerky sticks are finished in a few days; medium diameter forms, such as pepperoni may take a little longer; and the larger diameter salami can take a month.

12.2.4 Slicing and Packaging

Lunch meats for retail consumer use are sliced and packaged automatically. This is a potential source of contamination because the products are exposed to air and the knives are cutting through the surface across the interior. *Listeria monocytogenes* is a special concern because a very low dose can cause serious illness; it is tolerant of low temperatures and it is commonly found in meat plants. There have been numerous outbreaks and food recalls because of *Listeria* on ready-to-eat processed meats. One solution is heat pasteurization after packaging, using carefully controlled hot water, but there is concern about cooking the meat and about the time to achieve effective heating.

A promising approach is surface pasteurization of the meat using a steam flush just before the package is closed. The system has been demonstrated on frankfurters (Kozempel et al. 2003). Irradiation of chub-packed ground meat is effective against *E. coli* O157:H7 but is not yet widely practiced. Fresh sausages are kept refrigerated and must be cooked by the consumer by boiling in water, frying, grilling, or baking.

12.3 Sanitary Design of Meat Processing Plants

Because of their vulnerability to contamination, meat products must be manufactured under sanitary conditions, with buildings and equipment properly designed and operated. Other sources treat more thoroughly the issues of proper food plant design (Clark 2009, Lopez-Gomez and Barbosa-Canovas 2005, and many others). It is worthwhile including here in Table 12.1 a summary of the sanitary design principles developed by a special committee of the American Meat Institute.

Table 12.1 Facility Design Task Force (FDTF) mission

Establish sanitary design principles for the design, construction, and renovation of food processing facilities to reduce food safety hazards

12.3.1 Final FDTF Principles and Expanded Definitions

Principle 1: Distinct Hygienic Zones Established in the Facility

Maintain strict physical separations that reduce the likelihood of transfer of hazards from one area of the plant, or from one process, to another area of the plant, or process, respectively. Facilitate necessary storage and management of equipment, waste, and temporary clothing to reduce the likelihood of transfer of hazards.

Principle 2: Personnel and Material Flows Controlled to Reduce Hazards

Establish traffic and process flows that control movement of production workers, managers, visitors, QA staff, sanitation and maintenance personnel, products, ingredients, rework, and packaging materials to reduce food safety risks.

Principle 3. Water Accumulation Controlled Inside Facility

Design and construct a building system (floors, walls, ceilings, and supporting infrastructure) that prevents the development and accumulation of water. Ensure that all water positively drains from the process area and that these areas will dry during the allotted time frames.

Principle 4. Room Temperature and Humidity Controlled

Control room temperature and humidity to facilitate control of microbial growth. Keeping process areas cold and dry will reduce the likelihood of growth of potential food borne pathogens. Ensure that the HVAC/refrigeration systems serving process areas will maintain specified room temperatures and control room air dew point to prevent condensation. Ensure that control systems include a cleanup purge cycle (heated air make-up and exhaust) to manage fog during sanitation and to dry out the room after sanitation.

Principle 5. Room Air Flow and Room Air Quality Controlled

Design, install, and maintain HVAC/refrigeration systems serving process areas to ensure air flow will be from more clean to less clean areas, adequately filter air to control contaminants, provide outdoor makeup air to maintain specified airflow, minimize condensation on exposed surfaces, and capture high concentrations of heat, moisture, and particulates at their source.

Principle 6. Site Elements Facilitate Sanitary Conditions

Provide site elements such as exterior grounds, lighting, grading, and water management systems to facilitate sanitary conditions for the site. Control access to and from the site.

Principle 7. Building Envelope Facilitates Sanitary Conditions

Design and construct all openings in the building envelope (doors, louvers, fans, and utility penetrations) so that insects and rodents have no harborage around the building perimeter, easy route into the facility, or harborage inside the building. Design and construct envelope components to enable easy cleaning and inspection.

Principle 8. Interior Spatial Design Promotes Sanitation

Provide interior spatial design that enables cleaning, sanitation, and maintenance of building components and processing equipment.

Principle 9. Building Components and Construction Facilitate Sanitary Conditions

Design building components to prevent harborage points, ensuring sealed joints and the absence of voids. Facilitate sanitation by using durable materials and isolating utilities with interstitial spaces and standoffs.

Principle 10. Utility Systems Designed to Prevent Contamination

Design and install utility systems to prevent the introduction of food safety hazards by providing surfaces that are cleanable to a microbiological level, using appropriate construction materials, providing access for cleaning, inspection, and maintenance, preventing water collection points, and preventing niches and harborage points.

Principle 11. Sanitation Integrated into Facility Design

Provide proper sanitation systems to eliminate the chemical, physical, and microbiological hazards existing in a food plant environment.

May 4, 2004

While primarily intended to guide design of meat processing plants, these principles can be applied to other food plants as well. One aspect to note is that meat plants are commonly cooled to below 50°F, which creates a high relative humidity atmosphere. Thus some of the principles address the consequences of using hot water to clean equipment in such an environment. Fog and condensation will form and create hazards of low visibility and slippery floors.

Meat plants are regulated by the United States Department of Agriculture (USDA), which has traditionally required that designs be approved in advance. This requirement can be contrasted with the approach of the Food and Drug Administration (FDA), which regulates all other areas of the food industry (with some minor exceptions – eggs and seafood). USDA wants designs to have logical material and people flow, plenty of sanitation facilities such as hand washing stations, and materials of construction that are easily cleanable and can be inspected. Equipment used in meat processing plants must be on a list of approved equipment and must meet the same requirements of being able to be easily cleaned and inspected.

12.4 Examples

1. A new process and plant experience changes.

   Breakfast sausage is a ground fresh pork mixture with characteristic spices, including sage, salt, and pepper. It may be formed as links, patties, or chubs and the links may be cased with natural or collagen casing or have no casing. In the last case, the mixture is extruded into short lengths and then frozen. Some extruded pieces are partially cooked before freezing to make brown-and-serve sausages. In general, sausages are usually made from trimmings and less valuable cuts of pork, but some premium products are "whole hog," meaning they use all of the animal, including such primal cuts as hams, loins, and bellies. Often, whole hog sausage is made from large breeding sows that have become less productive. Such animals may reach 800 lb, while the normal butcher hog is about 200 lb at slaughter.

In the early 1980s, a large meat processor decided to convert an existing building to manufacture whole hog sausage using hot deboning and very rapid processing into chubs so as to extend shelf life. As previously mentioned, the natural microflora on meat is adapted to the animal's body temperature of about 100°F. If the meat is removed, ground, stuffed into a protective package, and rapidly cooled, the microbes adapted to body temperature tend to die off and psychrophilic microbes (those that grow best at low temperatures) do not have a chance to proliferate. The process concept was called, somewhat inaccurately, aseptic sausage. (Inaccurately because, strictly speaking, the meat was not sterile, as "aseptic" implies, but rather had a modified microbial ecology.) As compared with a conventional small meat packing plant, the proposed plant would have several novel features.

As an exercise, before reading further, list some of these features.

The plant was to handle about 80 large hogs per day or about 64,000 lb of meat on the hoof. Yields from hogs are about 50% of live weight, so the expectation was for about 32,000 lb of sausage, less any loss and plus added ingredients, such as spices. At the time, the standard composition for fresh sausage was about 55% fat and it was expected that the hogs might very well have a higher fat content because of their age and size. One of the first questions raised by the relatively young project manager (guess who that might have been) was what to do with the presumably excess fat. The answer at the time was that everything would go into the product. This violated no regulations but might lead to inconsistent product quality.

Ponder or discuss the ethical and business consequences of this issue and possible actions by the various parties. Besides the corporate project manager, there was a division engineering manager, an outside consulting engineering firm, and the proposed plant manager.

Some of the unique features of the plant included the following:

- Stronger than normal conveyor chain because of the anticipated weight of the hogs
- Absence of chillers
- On-rail cutting into quarters
- Use of a skinning machine provided by a shoe company that would take the hides
- Boning tables for hot meat deboning
- Grinding and stuffing equipment for warm meat
- Glycol bath cooler for chubs.

Thinking about the glycol bath, estimate the cooling load required. What information is needed for the calculation? Assume the plant operates 8 h/day and that chubs weigh 1 lb each. Typically, fresh breakfast sausage chubs are about 3 in. in diameter. The goal was to reach a center temperature of 35°F in about 1 h. How big would the bath be? The cooling rate calculation uses the published solution for unsteady conduction heat transfer in a cylinder.

Such cooling baths are fairly common in food processing. Another application is for cooling hot filled bags of fruit purees. Such bags might have 10–20 lb of product. In considering the cooling load, it is important to distinguish between the load due to the product and the load required to cool the bath down initially. In one case, involving hot-filled fruit, a plant was wearing out refrigeration compressors on such a bath because they were not pre-cooling the bath before introducing product. It was suggested to them that the bath be turned on several hours before starting production. How many hours might be required if the compressor were properly sized for the product load? Assume the conditions estimated for the sausage case. Would it make sense to just leave the compressor running all the time instead of turning it off at the end of each day and starting early the next day? Assume the bath is reasonably well insulated and is in a refrigerated environment – < 50°F is common for meat plants. What if the plant is not cooled and is in a warm part of the country?

Some other lessons were taught by the experience with the sausage plant. Once it was operating, it turned out that most of the raw materials were ordinary butcher hogs of about 200 lb. It took more of these to make the same amount of sausage, so more people were required than had been assumed. They were also leaner, so the average composition of the sausage changed. This was consistent with consumer trends toward less fat in the diet, but significantly changed the characteristics of the product.

Plant operators started cutting out valuable cuts such as loins, which fetched much higher prices than did sausage (loins can be used to make Canadian-style bacon), but there were no coolers in which to store separate products. Sometimes they packed the premium cuts in cardboard boxes and added dry ice to cool them down. Ribs and bellies were also tempting to harvest, but doing so reduced the lean component of the sausage and also required designated labor. It is questionable whether the extra value was worth more than the extra cost involved.

Eventually, a competitor acquired the company and the small sausage plant was closed.

2. Hazards of optimizing.

The values of different cuts of meat vary daily. Typically prices are published in newsletters circulated within the industry. A meat processor decided to try to optimize the value realized from hogs by correlating the composition of hogs with characteristics that could be measured. Specifically, they wanted to adjust how a cut was made that determined whether meat became part of a loin or part of a ham. Measurements were made of the length, weight, and fat thickness of many hogs and correlated statistically with the yield of certain cuts. This amounted to a mathematical model of a hog. Keep in mind that this effort occurred in the time of large, slow mainframe computers.

In practice, each hog was identified and its easily measured characteristics were taken and entered through an early voice recognition system by staff positioned on the line wearing headsets. When the carcass got to the station where the critical cut was to be made, the computer, which had crunched the numbers in the meantime, adjusted the position of a power saw blade. The operator of the saw was instructed always to fix a beam of light on a certain joint and let the computer control where the cut was made. Previously, he or she would have made the cut as they chose or were generally instructed.

Investment in gathering the data, analyzing it, and constructing the system was substantial, especially given the low margins of meat packing and the general lack of sophistication at the time in the industry. Did it work?

Well, yes and no. It worked for a time, or so it was claimed. One challenge is to ask how would one know whether it worked? Data had been gathered for average profit per hog for a base line period of time, before the system was implemented. It was assumed that the critical cut had been made randomly. Complicating the analysis is the fact that markets for pork change seasonally and for other reasons, so one time period is hard to compare with another. In any event, the assertion was made that after the system was implemented, average profit increased significantly.

A few years later, a similar comparison was made, only this time there did not seem to be any difference between profit with or without the system. What happened? It developed that the correlating equations had been almost continuously changed as additional data were collected. This meant that the underlying model was not stable. One might have thought that more data would cause the model to converge, but for a live animal that itself was undergoing significant changes due to breeding, nutritional advances, and consumer preferences, this was not true. Even if the characteristics of the hog had been stable, constantly adding new data to the correlation caused it to vary until it became a clumsy random number generator. The system was shut down. Could it have been salvaged?

Today, of course, computing power has increased greatly. There have been advances in instrumentation that enable fast measurements of more relevant characteristics. Hogs can be purchased individually instead of in lots and growers can be rewarded for providing more profitable animals. Robots under computer control perform some tasks in meat packing. It is possible that fast algorithms to optimize discretionary actions can be applied.

One universal lesson is to scrutinize correlations and mathematical models very carefully. Sometimes the data are obtained over such a short range that variations are little more than noise.

3. Intellectual property.

Pizza toppings made with meat are among the more popular and valuable products of the meat industry. Pepperoni and Italian sausage are two examples. In the early 1980s, a small manufacturer of Italian sausage, whose products were used by many pizzerias, developed a way to make pre-cooked Italian sausage pizza topping that looked as if it had been torn by hand from a chub. Pre-cooking was desirable to avoid release of free fat on the surface of the pizza and to shorten the bake time. Among the large pizza chains, fast service is the constant objective.

There are many ways the challenge could have been approached, including cooking bits of sausage in water or oil and cooking larger amounts before breaking them up. Some approaches resulted in pieces that were too uniform ("looked like rabbit droppings") while others had low yields or high amounts of fines. Eventually, the company developed and patented a process in which fresh Italian pork sausage was cooked in large diameter, perforated casings, and then extruded through a die plate that had irregular shaped holes. Italian sausage, by the way, is characterized by its spices, which must include fennel as well as salt and black pepper. The company obtained a patent on its process and equipment design.

This is an occasion to discuss patents and trade secrets. Patents are a legal grant by the federal government to an inventor that allows him or her to prevent others from using the invention in question without permission. The grant, which lasts 17 years in most cases, is in return for the inventor disclosing the invention to the public so as to advance science. After the patent expires, the invention is available to anyone. Because invention claims often overlap, obtaining a patent does not always mean an inventor can practice his or her own invention because some elements may be covered by another person's patent. Inventions that are patentable must be non-obvious, useful, and reduced to practice within 1 year of filing an application for a patent. Reduction to practice includes filing a patent application. The invention cannot have been used commercially more than 1 year before filing. In the United States, the first inventor to prove he or she conceived the concept is given priority. In other countries, the first person to file is presumed to have priority.

Trade secrets are intellectual property (IP) that are often defined by state law. Typically a trade secret must be valuable to the owner, must be protected by efforts to keep it secret, and should not easily be reproduced by some one else without specific information.

In the case of the new Italian sausage pizza topping, the product was a success because it addressed the needs of the customers. A large pizza chain ordered large quantities of the topping, but insisted that the small manufacturer license its process to other suppliers so that the chain had alternate sources of its ingredient. This is a common business practice. The small company licensed the process to competitors, in some cases sold equipment using its designs, and collected royalties. Royalties are negotiated fees paid by users of a patent to the inventor.

It is common for patents to be challenged in federal court, usually on the grounds that the patent should not have been issued in the first place because it did not meet the criteria of being non-obvious or for some other deficiency. Challenges often come from competitors or licensees who want to practice the invention without paying fees. In this case, the challenge was on the basis that product using the invention had been sold commercially more than 1 year before the application for a patent was filed. The inventor's defense was that such sales were in the normal course of development, which often is true, but the court ruled otherwise. With the patent ruled invalid, licensees stopped paying royalties and the large customer canceled its purchase agreements. Instead, the customer turned to a large meat packer and helped that company to produce the pizza topping at a lower cost because of their scale and integrated operation.

The small company sued the pizza company and the large competitor for trade secret misappropriation. They could not sue for patent infringement because the patent had been declared invalid. Trade secret misappropriation is somewhat more difficult to prove because often it is an afterthought in a legal proceeding that usually starts with patent infringement. Because trade secrets are specifically not disclosed widely, they may not be well documented. Trade secrets often exist with or without a valid patent. A patent discloses the widest possible range of operations in order to cover as much territory as possible. Negotiating the specific claims in a patent is a time consuming exercise between the inventor and his attorney and the patent examiner at the Patent and Trademarks Office (PTO) in Washington. The inventor is required to disclose his or her preferred implementation of the patent, but it is common that there are still "tricks" involved that are learned by experience and that are valid trade secrets. A good example of a true trade secret is the specific formula or recipe for a food product.

As an exercise, list some of the trade secrets from your own company or from another source. (The proprietary details are not necessary, just the names or categories.)

Over the course of more than 13 years, the small meat company reached modest settlements with several of its competitors, but its cases against the largest competitor and its customer dragged on. At one point, the case against the customer was denied because of a jurisdictional dispute, but then was reinstated. The case against the large competitor went to trial in federal district court and the small company won a substantial award for damages, which was largely upheld on appeal. The testimony that almost certainly decided the case in the eyes of the jury came from a former employee of the small company who had been recruited over a long distance for a job at the larger company, after he had left the small company. He had helped practice the trade secrets in the small company and was specifically asked to help the larger company establish its version of the process. Once the new operation was running smoothly, he was discharged.

It is well established that employees are entitled to use their experience and expertise in their careers, but they are also often bound by non-compete and non-disclosure agreements. Employers who deliberately seek out employees or former employees of competitors with knowledge that might include protected trade secrets are risking the accusation of misappropriation, which can be expensive.

Another element of the case was the testimony of an engineering expert witness who was permitted to visit and observe the operations of both the plaintiff and, under a court order, the defendant. While there were superficial differences in the way the two plants were operated, the expert concluded that they were substantially the same and that there was little likelihood that the defendant would have developed such a similar process entirely on their own without access to specific information, which the other testimony showed they had, in fact, received.

If you are an employer, do you have procedures to protect your own trade secrets? Do you know what they are? Do you have procedures to protect yourself against the accusation of misappropriation of other's IP?

In several cases, companies have hired outside experts to help them screen ideas that are constantly being offered to large companies. The odds are great that the companies already are working on most of the ideas, but they do not want to pass up the chance that something worthwhile is being offered. To protect themselves, they instruct the experts simply to advise whether or not to pursue a relationship with the outside party. The outside party, of course, must trust and approve the experts, who are usually paid by the company receiving the solicitation.

There is developing a greater appreciation of open innovation, which is a more proactive exercise of looking for useful ideas outside of the firm. There exist organizations that facilitate the identification of sources of ideas and presentation of specific challenges to outside experts and organizations. Examples include NineSigma, YourEncore, and Innocentive, all of which operate over the Internet.

12.5 Lessons

1. The principles of sanitary meat plant design can be applied to almost any food plant.
2. It is necessary to anticipate possible changes in operation and requirements in any food plant design because circumstances do change. One of the best ways is to leave extra space.
3. Mathematical models must be developed with data from a wide range in order to be robust.
4. Intellectual property is valuable and must be identified and protected.
5. No single company has a monopoly on good ideas or talent. It is becoming accepted to look outside the organization for inspiration.

Chapter 13
Non-thermal Processing

Non-thermal processing refers to relatively young technologies that use mechanisms other than conventional heating to reduce or eliminate microorganisms that might be harmful or cause spoilage. There is a large, active division of the Institute of Food Technologists (IFT) devoted to the topic and the various processes have been the subject of much research and publication. This chapter will briefly summarize some of the technologies and discuss various applications (if any) and what some of the practical and theoretical issues are. Some of this material appeared in a different form in the Processing column of *Food Technology* beginning in 2002.

Thermal processing is still the major technique for shelf-stable food preservation, but other technologies are finding applications. There is a general movement in food processing away from high-heat treatment and deep-freezing toward milder treatments, resulting in refrigerated foods with less cooked flavors. There always is the need to kill pathogens, but there is also a demand for "clean" labels, meaning a preference for few, if any, chemical additives and preservatives.

Traditional thermal processing was focused on killing spores of *Clostridium botulinum* in low-acid foods. Refrigerated foods, which often receive little heat treatment, can be contaminated with less-heat-resistant species such as *Listeria, Salmonella,* and *Escherichia coli*. Non-thermal techniques, such as irradiation and high hydrostatic pressure, can destroy these organisms with little damage to the food.

13.1 Evaluating Non-thermal Processes

Older preservation processes such as acidification, dehydration, and antimicrobials have a substantial impact on food composition and sensory attributes. They are bacteriostatic, not bactericidal, i.e., they inhibit the growth of microorganisms but do not necessarily kill them. In contrast, non-traditional processes are driven by the desire to have minimal impact on sensory properties, can be complex, and can induce physical changes, including heating, since many involve applying novel forms of energy. They may either inhibit or kill microbes, depending on the circumstances. In any case, new preservation processes must be approved by the Food and Drug Administration (FDA).

J.P. Clark, *Case Studies in Food Engineering*, Food Engineering Series,
DOI 10.1007/978-1-4419-0420-1_13, © Springer Science+Business Media, LLC 2009

Food and Drug Administration's approach to risk management has changed in recent years. The traditional standard was to achieve a risk level "as low as reasonably achievable" (ALARA). "Reasonable" was in the eye of the beholder, a matter of judgment. It was common to assume a worst case, e.g., high initial levels of contamination, overfilling of a container, and higher viscosity than target. Such assumptions led in the case of standard practice for canning to targets of 12 decimal (12D) reductions in *C. botulinum* spores. Such thermal processes typically result in overcooking foods and reduced levels of vitamins and color.

The new approach at FDA relies on setting a public health goal and concentrating on achieving that goal rather than specifying how it is achieved. There are two significant elements on which the agency focuses: validation and verification. Validation demonstrates that a given goal can be achieved. Verification assures that the goal actually is achieved. A Food Safety Objective (FSO) is established, which, expressed as decimal reductions, must be greater than the sum of the initial level of contamination minus the reduction achieved, plus any increase or recontamination. The critical question, of course, is what the "appropriate level of protection" (ALOP) is?

For instance, even if there is 95% confidence of achieving some target, if there are three steps, each with that level of confidence, there could be a 14–15% chance of failure. Thus, validation implies not that a process cannot fail, but rather that the rate of failure is within acceptable limits. This is a highly significant point – approved preservation processes may have a finite, though small, probability of failure, even when operated correctly.

13.1.1 Validating a Process

The key to process validation is a process capability study in which the results are typically presented as a control chart, showing a process outcome as a function of time. The process outcome would typically be the decimal reduction actually achieved, for instance. One practical issue is how to detect extremely low levels of microorganisms in foods.

Thus, process capability studies typically involve challenges, often using surrogate organisms added to a food in such levels that even after significant reductions they can still be detected. An important principle in such studies is that the process should be driven to failure, typically by using extremes of process conditions, to understand the ruggedness of the process. FDA wants to see the data from such a well-designed study for any new technology.

As a recent example, the HACCP guidance for fresh fruit juices specifies a 5D reduction in pathogens. The goal was risk reduction, not elimination, and the agency hoped to preserve at least some distinctive fresh flavor in the products. Exceptions to the rule requiring a mild thermal process are pasteurized shelf-stable juices, which undergo a more severe process anyway, and citrus juices made from surface treated, handpicked sound fruit, which are considered to be essentially sterile, provided that the skin is washed with hot water and sanitizer.

A major challenge for new processes is knowledge of the kinetics of bacterial destruction, so that conditions can be simulated and results correlated. Kinetics are understood for irradiation and are beginning to be understood for high pressure, but other processes require continuing research in this area. The high-pressure model applies to vegetative cells but not, so far, to spores.

By law, some processes require premarket approval because they have been defined as additives and so are regulated as such. Examples are all radiation processes (not just ionizing radiation) and antimicrobials that are not generally recognized as safe (GRAS). High pressure is not defined as an additive, but how FDA will consider any filings for approval will depend on whether the treatment has a lasting effect.

An important element of any study submitted to FDA is evidence that diversity has been considered, meaning potential biological variation in target microbes, variations due to equipment, geographical variations (e.g., environmental conditions), and variations in the food itself. Changes of just 0.1 in pH units can matter in food safety. The significant data that must be measured and submitted include the initial load, the reduction achieved by the process, and the increase or recontamination that could occur, all measured over a variety of conditions and recognizing the potential for variation. The FSO must be specified and defended, using, for instance, the limit of detection that may be appropriate.

There is also a complex debate over the description of food as "fresh." Many non-thermal processes claim to retain "freshness" in some sense. FDA is concerned about the legal and scientific bases for such claims.

Finally, FDA wants surrogate organisms to be more resistant to a process than the target pathogens. FDA also wants at least one test with "real" pathogens, which itself causes anxiety in researchers, who are conscious of the dangers of such microbes. For that reason, most process capability studies are performed in pilot plants rather than in production facilities.

13.1.2 Determining Kinetics

There are several candidate kinetic models for new processes, in particular, several modifications to the traditional first-order rate equation. Thermal processes have traditionally been described with a first-order differential rate equation. In integrated form, this equation shows a linear decline in relative numbers of microbes with time at constant temperature when plotted on semi-logarithmic graph paper. This gives rise to the convention of decimal reductions in describing processes. For some new processes, it is possible that this reliable model does not apply, so the mathematics and experimentation can become more complex.

For high pressure, the rate constant may be a function of pressure and most high-pressure processes also involve some adiabatic heating, so temperature almost certainly changes. Adiabatic heating means an energy change created within the product as distinguished from heating by an outside source. Spores may be injured by decompression in addition to whatever damage heating under pressure may cause. It can be difficult to distinguish between injured and permanently inactivated cells.

These complications for high pressure can then be compared with issues that arise in other processes, such as microwave and radiofrequency heating, ohmic and induction heating, pulsed electric fields, high-voltage arc discharge, pulsed light, oscillating magnetic fields, and others. The basic question in every case is what is the required time at given process conditions to achieve a given FSO. Some of the new processes mentioned will be described later. Collectively, they are various means of delivering energy to a food and the microbes it contains so as to kill or inhibit the microbes that cause disease and spoilage.

13.1.3 Finding the Data

There is a database of microbial responses in food environments assembled by researchers at the USDA Eastern Regional Research Center (ERRC) in Wynd-moor, PA. Called *ComBase* and available at www.combase.cc, it has more than 32,000 records of growth and survival for microbes in foods. There are data for 28,000 pathogens and 4,000 spoilage microorganisms. Access is free. *ComBase* data can be integrated with ERRC's Pathogen Modeling Program, found at www.arserrc.gov/mfs/pathogen.htm. The point is that this information, previously scattered in many forms throughout the literature, is now being collected and presented in a convenient form.

13.2 High Hydrostatic Pressure

It has been known that very high hydrostatic pressure can kill vegetative microbes, but the systematic application to food preservation is relatively recent. The pressures involved, 60,000 psi and higher, require specially designed chambers and an economic method of creating the high pressure. So far, high pressure processing has been a batch operation. Current commercial products processed with high pressure include guacamole, ready-to-eat meat, and fruit juices.

High pressure does not inactivate enzymes, which can cause various negative effects, such as discoloration, bitter flavors, and softening. Some mild heating is required to inactivate enzymes. A specific example is preparation of a fruit mixture, such as might be used in fruit salad. Using deflavored lemon juice lowers the pH, though other acids could also be used. The mixture is heated at 70°C for about 10 min and then subjected to 85,000 psi. The result is almost fresh tasting.

A variation is called pressure-assisted thermal treatment, in which food is heated to about 90°C and then pressurized to about 100,000 psi. The adiabatic heating raises the temperature to about 121°C, hot enough to kill spores of *C. botulinum*. This permits treatment of low-acid foods, and the heating inactivates enzymes.

The advantage of the process is that when pressure is released, cooling is very quick. In conventional thermal processing, cooling can be slow, and much of the damage to flavor and texture occurs then.

13.2.1 Equipment

Because of the high pressures, the process is conducted in sturdy vessels and usually as a cyclical batch operation. The producers of commercial equipment include Avure, Kent, WA; Elmhurst Systems, Albany, NY; Engineered Pressure Systems, Inc., Haverhill, MA; Stansted Fluid Power, Essex, UK; and NC Hyperbaric, Burgos, Spain. Avure's equipment grew out of the technologies of its parent companies, ABB Pressure Systems and Flow International, in high-pressure compaction of ceramics, powder metal, diamonds, and water-jet cutting, in which a thin stream of water is pumped to about 50,000 psi up to 1,800 times a minute, using pistons. These experiences gave Avure the know-how to manufacture large HPP systems with the duty cycles required for the food industry.

Elmhurst builds its high-pressure units from surplus cannon barrels. Its first commercial installation was a reliability demonstration. The goal was to demonstrate 15 cycles/h for 20 h over multiple days. The cycling stressed valves, piping, and instruments, as well as the pressure vessels.

Most HPP vessels use an external frame or yoke and usually have a non-threaded closure. Because water and foods compress under pressure, pumps must supply up to 16% additional water during a cycle. This water is usually captured when pressure is released and then screened and re-used. The temperature of the water is usually controlled. There is a substantial adiabatic heat of compression which generally raises temperature about 3°C for each 100 MPa (mega Pascal, about 10 atmospheres) when water is the pressure transferring medium. This temperature change, however, is reversed upon decompression.

13.2.2 Mechanism

High pressure is believed to preserve foods by inactivating vegetative cells. High pressure alone does not appear to kill significant populations of spores, but it does appear to induce germination. Spores appear to be inactivated by combinations of both pressure and heat. Spores of *Bacillus amyloliquefaciens* are possible surrogates for pathogenic *Clostridium* and *Bacillus* species based on resistance to pressure and temperature.

Neither pulsed electric fields (PEF) nor HPP appear to have much effect on most enzymes that affect quality of foods so enzymes may then determine the acceptable shelf life of non-thermally treated foods. Researchers have found that high pressure might even stabilize some enzymes.

Several hydrolytic reactions catalyzed by enzymes proceeded more rapidly at elevated pressures than at atmospheric pressure and the same temperatures. The research showed that the effect was due to retarding of the inactivation of the enzymes by heat, permitting the reaction to occur at high temperature, which increased the reaction rate. Pressure actually delayed the reaction, but the ability to use higher temperature more than compensated for the delay. The net effect was an increase of 20–100% in substrate conversion at optimum pressures and temperatures, which varied for each enzyme and substrate combination. Optimum pressures were 106–318 MPa, and optimum temperatures were 55–84°C. These results suggest that the mechanism of HPP is not inactivation of enzymes. Other research is inconclusive on whether there are physical changes to cells under HPP.

Some researchers have found no morphological changes in pressure-treated spores of *Encephalitozoon cuniculi* in apple cider. In contrast to noting no morphological change, others observed significant physical damage by HPP to spores of *Bacillus subtilis* treated in milk. HPP has been combined with use of the bacteriocin nisin, a natural fermentation product that is a mild antibiotic approved for certain foods by the FDA.

Pressure and temperature have effects on the pH of various buffers. Some buffers are less sensitive than others to pressure, but since there is almost always a temperature change due to high pressure, the sensitivity to temperature is also a factor. Perhaps a pressure-induced pH change in foods contributes to inactivation in certain foods. Another hint as to a possible mechanism for HPP is the effect of high pressure on dairy proteins and products. Some of the effects of HPP on milk include a reduced size of micelles (the small aggregates of protein and calcium suspended in milk), increased calcium solubility, denatured whey protein, and a change in color. HPP also appears to accelerate ripening of Mozzarella cheese. HPP might be applied to milk as much for its impact on functional properties as for its preservative effect. There are no HPP dairy products yet, but one possibility is cheese made from raw milk, which is popular, especially in France, but which carries a risk of *Listeria monocytogenes*. HPP eliminated *L. monocytogenes* from soft cheese made from inoculated raw milk. Prototype extended shelf-life fruit yogurt products for use on the NASA space shuttle have been prepared.

13.2.3 Other Applications

Inactivation of *E. coli* O157:H7 in tomato juice and liquid whole egg has been studied. Liquid whole egg, of course, is vulnerable to heat and was found to coagulate at the highest pressure and temperature used (600 MPa, 50°C). Treatments were for relatively long times (up to 60 min) at near room temperature. Cycling the pressure up to four times for a total exposure of 40 min was more effective than using continuous pressure for the same time. This raises the possibility that pressure changes have an influence on inactivation. Reductions of 3–5 log cycles were achieved.

The inactivation of four types of coliphages was studied to evaluate their potential as human enteric viral surrogates. Coliphages are microorganisms that infect bacteria. The coliphages differed from each other in their sensitivity to high pressure at room temperature. Pressures up to 600 MPa for 5 min were required to get significant reductions in the most-resistant phages.

HPP actually seemed to protect cells from PEF under some conditions. However, at other conditions, there was a synergistic effect, suggesting that both processes may affect cell membranes.

Short exposure times of microbes in beverages gave significant reductions (7–9 log cycles) in milk and fruit juices at times as low as 10 s. Some organisms required up to 60 s and temperatures of 75°C for inactivation, but these are still fast processes compared to many others.

13.2.4 What's Next for HPP?

The most successful pressure-treated food product, so far, is guacamole. However, red and poultry deli meats have been successfully introduced into many retail stores. All food examples to date have been refrigerated foods. The lack of success against spores has meant that HPP cannot yet produce a shelf-stable low-acid food. Most foods now are treated in flexible or semi-rigid packages, in batches, and in cycles measured in double-digit minutes. There is interest in treating fluids. One scheme pumps the fluid into a holding vessel and uses several such vessels to approximate continuous flow. Another scheme for fluids would use large bags, such as those used for bulk aseptic storage. This would still be a batch process and would require a larger chamber than has been used so far.

Another area of application for HPP is in treating shellfish. Oysters are simultaneously shucked and pasteurized without fully cooking, while lobster meat is removed from its shell by HPP.

13.2.5 Effects of HPP

The following are some of the effects of HPP that have been noted:

- *Protein unfolding.* High pressure can unfold proteins in a somewhat controlled manner as a function of pressure and temperature. There is a company, pressurebiosciences.com, that makes high-pressure equipment just to unfold proteins. Shucking of shellfish (oysters, lobster, crab, etc.) is a popular application of HPP in the range of 40,000 psi.
- *Cell wall opening.* Researchers have championed HPP, PEF, and ultrasonics for opening cells to promote quick release of useful secondary metabolites. A mundane example would be to increase the extraction of sugar from sugar beets.
- *Setting of starches.* HPP will set starches without heat, yielding a raw starch gel.
- *Pressure freezing.* This process can be used to produce very small ice crystals. Foods are cooled to −20°C under pressure. Under a pressure of 27,000 psi, the water remains in liquid form because the high pressure counters the expansion of water when it tries to freeze during cooling. When the pressure is released, small

ice crystals can form throughout the food mass to help freezing at atmospheric pressure. Also, foods can be thawed quickly using pressure thawing by reversing the process, i.e., applying high pressure to frozen foods and then warming. The phase transition from ice to liquid water is almost instantaneous.

- *Diffusion.* Attempts have been made to use HPP to promote rapid diffusion of water into beans and other foods for rapid hydration.
- *Setting large wheels of cheese.* HPP can direct-set Cheddar cheese curd, ensuring uniform salt distribution and rapid processing.
- *Pressure blanching.* Some plant and animal enzymes are unstable enough to be inactivated to a useful low level by high pressure. Avocado is the best example, in which browning is prevented by HPP. HPP has allowed the refrigerated distribution of thousands of tons of guacamole – this may be the largest single application of HPP so far.
- *Enzyme activation.* Pressures in the range of 15,000 psi activate proteases and help tenderize meat prior to rigor.

13.2.6 Dairy Applications of HPP

Notably absent from the list of food products for which HPP has been commercialized are dairy products. While HPP-processed meat, juice, fruit, and shellfish have all been launched around the world in recent years, dairy products have not been among those commodities. This is not because HPP has no interesting or relevant effects on milk or dairy products; indeed, the opposite is the case. The effects of HPP on milk on dairy systems are complex, unique, and perhaps quite technologically significant. The argument may be made that it is precisely the complexity of the effects of HPP on milk which has hindered the commercial application, as scientists and technologists grapple with the findings of research in this area. A reflection of the state of dairy HPP research is the fact that far more papers have appeared in the scientific literature on the effects of HPP on milk and milk products than on almost any other commodity group.

One of the more interesting phenomena is the selectivity of HPP inactivation. Often, HPP will kill or inactivate spoilage and pathogenic organisms without seriously affecting probiotic or desirable cultures in products such as yogurt and other "healthy" dairy-based products, valued for their live cultures. HPP causes microbial inactivation in milk and many other food products, but spores remain a problem for low-temperature inactivation. It seems unlikely that HPP will become economically attractive as a milk preservation technology, compared to well established processes such as pasteurization and ultra-high temperature (UHT) treatment for extended-shelf-life milk.

An important factor in the inactivation of spores is their state of aggregation, which is measured optically. When this factor is incorporated into a model, the researchers can accurately correlate their data. Still, it is difficult to get an adequate reduction in spores in low-acid foods using HPP alone. Pressure-assisted heating, however, shows some promise.

HPP may have interesting applications for dairy products where heat treatment would be undesirable or infeasible (in common with many other products where HPP has found a unique niche, such as guacamole). For example, it may be possible to extend the shelf life of products such as yogurt or even quick-ripening cheeses (such as mold-ripened cheeses) by inactivation of microorganisms and enzymes, thereby arresting the metabolic reactions in the product, which would otherwise result in progressive deterioration of quality.

The concept of shelf life extension has been the first area where HPP dairy products have been commercialized. A soft cheese spread treated by HPP to inactivate microorganisms was launched in Spain, while a company in New Zealand has proposed use of HPP for preservation of probiotic yogurt and biologically active colostrum, which would lose their health-enhancing characteristics if heat-treated.

Early commercial results also suggested that HPP could greatly accelerate the ripening of cheese. More recent studies have shown that the magnitude of such effects in most cases is quite modest; but great effects on the functionality (e.g., melting and stretching properties) of Mozzarella cheese after HPP of the freshly made cheese have been reported. The process greatly accelerated the development of desirable properties for use in pizzas.

Another major area of potential interest involves treating milk with HPP and using it as a base for making dairy products because of the many unique effects of pressure on milk constituents. For example, HPP denatures whey proteins (as does heating) but, unlike any other process, it also affects the structure and properties of the casein micelles, resulting in either aggregation or disintegration of micelles, depending on the pressure, temperature, and pH. These changes have great implications for the production of dairy products from such milk, e.g., altering the rennet coagulation properties of milk and perhaps the texture and ripening of cheese. Researchers are studying the effects of HPP on cheese-making properties of milk for use in Cheddar cheese production. In addition, HPP also seems to have significant effects on acid coagulation properties of milk and if applied to ice cream mixes to have substantial effects on the final product characteristics.

13.3 Irradiation

Irradiation is a long-studied and reliable technique for treating a variety of foods with various effects. At low doses, below 1 kiloGray (kGy), potatoes are inhibited from sprouting, insects in grain are kept from reproducing, fruit is delayed in ripening, and mold on fruit is destroyed. At larger doses, 1–10 kGy, pathogens on poultry or meat are destroyed. And at the highest doses allowed, 10 kGy and above, spices and dehydrated vegetables are sterilized.

Three sources of radiation are approved: cobalt-60, electron beams, and X-rays. Cesium-137 is another potential source but is currently used for research, not production. Radioactive isotopes (cobalt-60 and cesium-137) and X-rays produce energy in the form of photons, while electron beams generate a beam of electrons. X-rays are created by first generating an electron beam and then directing it at a metal target from which penetrating photons are produced. The gamma rays from isotopes and the beams from an X-ray have different energies, but function in the same way.

Radiation doses are measured ultimately by sensitive calorimetry, i.e., measuring the temperature rise due to exposure for a given time. A Gray is defined as 1 J/kg, and 1 kGy is about 0.4 BTU/lb. At the typical doses used for food pasteurization, the temperature rise is about 1°. In practice, doses are measured by specially treated polymers, which undergo a color change in proportion to the dose received and which trace their calibration back to calorimetry.

Early research on irradiation of foods focused on sterilization for such applications as feeding astronauts in space and the military, but sterilizing doses can cause off-flavors in some foods. The process is still used for these special applications, but the large, civilian applications are for foods that are refrigerated or frozen and are pasteurized with lower doses than are required for shelf-stability.

The categories of irradiators refer to how the isotope source is stored and shielded. Category I is for small sources used in research. Category II uses air storage and air irradiation and is not used for food. Category III uses water shielding and irradiates under water. Category IV stores a source under water but moves it into air to expose a target. Typically, when a source is in air, it must have concrete shielding to protect surroundings from the radiation. Because of the potential hazard to workers, special training is required for using any radiation device, whether it be for medical use, as in treating cancer, irradiating medical devices, or irradiating food.

The Genesis irradiator, from Gray*Star Inc. in New Jersey, with its scheme of keeping the source under 20 ft of water, requires less shielding and is less complex than other designs. Products to be treated are loaded automatically into "bells," which use compressed air to displace water and are lowered through the water pool to a position next to the source. The source is also in air under the water, separated by a thin steel panel, which does not reduce the amount of radiation. The Genesis system occupies only 1,200 ft^{2}and can be built in about 6 months for about $1.25 million. Another $1.5 million is required for the initial load of cobalt-60. An owner of a Genesis unit obtains the source from firms that specialize in that business, such as MDS Nordion, a Canadian firm that specializes in technologies for nuclear medicine and irradiation, or Reviss, a British firm.

A significant operating cost of isotope irradiation is replacement of the cobalt-60, which decays at the rate of about 1% per month. To compensate for the decay, some fresh sources are typically added every year. Eventually, depleted sources are returned to the supplier and new ones installed.

The shielding water in irradiators must be kept very clean to avoid any corrosion of critical components and to enhance visibility, as operations such as replacing sources must be conducted remotely through the water.

The cobalt-60 sources are manufactured by exposing cobalt-59 rods in a nuclear reactor, where some of the atoms acquire an extra neutron to become the radioactive element cobalt-60. The cobalt-60 is then doubly encapsulated in stainless steel. Cobalt-60 is produced in power reactors, where the cobalt-59 serves to absorb neutrons while producing the valuable cobalt-60.

There are advantages and disadvantages to each source of radiation. Gamma rays are emitted in all directions and are generally limited to targets about 2 ft thick. This means that some energy can be wasted, depending on the design and the density of the targets. Typically, pallets of food in cartons are reconfigured automatically before passing through any irradiator, to assure uniform exposure without overexposure of the outer surface. Gamma rays can penetrate any product, but are affected by product density. They penetrate farther than do electron beams and about the same as X-rays. X-rays can be focused because they are generated in a beam, so while there is some energy inefficiency, typical pallets can be treated without reconfiguration. Electron beams have low penetration depth and so can only be used for thin targets, no more than about 3.6 in. thick, exposed from two sides. Typical electron-beam treatments are for single layers of packaged hamburger patties or for packaging material. Both electron beams and X-rays have the advantage of being turned on and off by a switch, if necessary. However, electron beams include a high vacuum chamber, which can take some time to be pumped down at start up.

Packaged ground beef is a typical application for electron beams. Electron beams are best with regularly shaped targets, while X-rays are better with thicker and irregular targets.

The suppliers of food irradiation equipment and services compete with one another, though costs of treatment are roughly equal, depending on utilization, energy costs (for those driven by electricity), and other costs. All foods pasteurized by irradiation carry the radura, an international symbol alerting the consumer that the food has been treated by irradiation. There has always been concern, heightened by the positions of some activists, that the public would resist buying irradiated foods. In fact, such foods have been successful when offered in the market. Efforts to generate protests and boycotts have failed, even in the home territory of some of the organizations opposed to irradiation.

The impetus for irradiating ground meat is concern about *E. coli*, *Listeria*, and *Salmonella*. The meat industry became more focused for a while on BSE (mad cow disease), which is not addressed by irradiation. Now, attention is shifting back to the pathogens, for some of which there is a zero-tolerance policy.

Irradiation is one of the better ways to ensure the safety of fresh and frozen meat and poultry. Post-processing contamination of ready-to-eat meat products, such as sausages, lunchmeats, and cooked ham, especially with *Listeria*, is a serious concern. A petition to the FDA to allow irradiation of ready-to-eat foods has been pending for several years. Irradiation is defined as an additive for regulatory purposes, even though it clearly is a process and very little is changed in the food. This status has slowed the approval process, but that approval may occur soon. This could create significant demand for new irradiation capacity.

Irradiation works by creating energetic electrons and free radicals in the product being treated. These are reactive species, which can interact with chemicals in cells and interrupt cell division. The same species can react with other components of food, but at the levels of radiation energy typically used there are rarely any noticeable effects on flavor, color, or nutritive value. Almost all spices, herbs, and dried vegetables imported into the United States have been irradiated, typically at 10 kGy, to sterilize and sanitize. Products used in foodservice or as components of other foods need not be labeled as irradiated.

The impact of radiation on quality is product specific and must be studied carefully. Once effective minimum and maximum levels are determined, the process is carefully controlled to stay within the limits. The response to the radiation is in proportion to the dose to which it is exposed.

A good reference on irradiation is by Komolprasert and Morehouse (2004). A consistent theme is the interaction among preservation, safety, and quality. For example, Chapter 7, by K.T. Rajkowshi and X. Fan, discusses irradiation of seeds intended for sprouting. The authors point out that a radiation dose high enough to give a 5-log reduction of some resistant pathogens would also kill the seed. Chapter 8, by B.S. Patil, discusses how irradiation improved functional components of fruits and vegetables. An example is increase in quercetin concentration in onions after relatively low doses. Quercetin is believed to have beneficial biological activity against cancer. Irradiation also seemed to increase flavanol concentration in citrus. This could be a positive side effect of irradiation to reduce insect pests traveling on fresh fruit.

Radiation is known to interact with polymers such as those used in food packaging. One effect can be chain scission, or the reduction in molecular weight of the polymer. Smaller fragments might have a greater tendency to migrate into food with which they are in contact. Radiation also can increase cross-linking, which increases molecular weight and would reduce migration of chemicals. The research in this area typically involves sensitive analysis of headspace volatiles before and after irradiation. It generally takes relatively high doses before significant amounts of new chemicals are detected.

Some packages used for bulk aseptic storage – large laminated film bags – are sterilized by irradiation before filling. Research has shown that there is no impact on seal strength even at very high doses.

The science to determine required doses and to control operations safely and economically is well established. Costs are reasonable, typically pennies per pound, given the significant health and quality benefits. Irradiation is safe for the users and consumers and is a valuable tool in the collection of techniques used to offer safe and nutritious food to the market. All irradiators operate under their own Hazard Analysis and Critical Control Point plan, in which dose, exposure time, and temperature are controlled. Given the demonstrated hazards of emerging pathogens such as *Listeria*, *E. coli*, and *Salmonella* and the ability of irradiation to eliminate them, it is likely that the radura will become a trusted symbol of safety on many packaged foods.

13.4 Pulsed Electric Fields

Pulsed electric fields imposes a strong electric field on a flowing fluid for a very short time. Above a critical field strength of about 15,000 V/cm, vegetative cells are killed. In practice, higher field strengths are used, up to about 35,000 V/cm for disinfection (destruction of bacteria, fungi, and other microbes). Lower-strength fields are used for static (non-flowing) PEF.

13.4.1 How PEF Works

When exposed to high electrical field pulses, cell membranes develop pores either by enlargement of existing pores or by creation of new ones. These pores may be permanent or temporary, depending on the condition of treatment. The pores increase membrane permeability, allowing loss of cell contents or intrusion of surrounding media, either of which can cause cell death. PEF has limited effects on spores and only appears to affect a few enzymes. Enzymes are important in juice processing because surviving enzymes can reduce pectin, which then can be less effective in keeping fruit particles suspended. Some sedimentation is common in fruit juices, but too much is unattractive. Some surviving enzymes may also enable discoloration and production of off-flavors.

Pulsed electric fields offers a 5-log reduction of most pathogens and is considered a pasteurization process, so products must be refrigerated. An important process consideration is prevention of post-process contamination, so filling is in an aseptic or clean room environment and containers must be cleaned and possibly sterilized.

The perforation of cell membranes caused by PEF also applies to fruit and vegetable cell walls, so a potentially beneficial side-effect of the process is improved extraction of juice from cells. This phenomenon is also applied in another promising use, concentration of sewage sludge, a suspension of live and dead cells and organic matter, which can be very hard to filter and concentrate. PEF, by killing live cells and reducing their ability to retain water, greatly improves filtration. Extraction of sugar from beets and starches from potatoes may also be improved by PEF.

13.4.2 Process Variables

Important process variables include the electric field, which can have various wave forms, strengths, and distribution in the treatment chamber; temperature; pressure; and time of exposure.

- *Electric field.* The electric field is generated by equipment similar to that used in radar. This has some consequences for cost and availability. The most typical equipment generates a short square wave and reverses polarity, in part to avoid erosion of electrodes. However, a bipolar generator costs about twice as much as a monopolar one. Other wave forms include exponential decay and sinusoidal. The sinusoidal form is somewhat easier to generate, as it uses equipment similar to common radio equipment, but it reaches its peak power only for an instant and so delivers less energy per cycle above the critical field strength than does a square wave.

- *Temperature.* There is some temperature rise across the PEF treatment chambers due to the delivery of electrical energy, but the preservative effect is primarily non-thermal. A typical temperature change is about 30°C for orange juice and lesser for apple juice. Processes typically operate at 35–50°C because it has been found that microorganisms are more tolerant at low temperatures.
- *Pressure.* Pressure is applied to inhibit the formation of air bubbles in which electrical arcing could occur with fields above 20,000 V/cm.
- *Time of exposure.* The field is cycled about 1,000 times/s, and the fluid is exposed to multiple pulses by passing it through several treatment chambers.

The treatment chambers are separated from each other by insulators, which may be ceramic or polymer. Each chamber has two electrodes, which are made from a conductor or semiconductor. Ohio State University holds patents on PEF treatment chamber and pulse generator designs.

Options for chamber design include parallel plates, co-axial, and co-field (the OSU approach). PEF may also be applied in a static (non-flowing) approach for foods such as meats. Electrical equipment components exposed to more than 35,000 V of potential must be immersed in oil for insulation and cooling. Issues in the equipment design include pulse coupling, switch design, high-voltage power supply, and over-current protection.

One supplier of the electrical equipment is Diversified Technologies, Inc. of Bedford, MA. DTI is primarily a defense contractor and acknowledges that designs for industrial food processing are only now emerging. So far, most of the PEF units that have been built are used in research and development, but one commercial application has received FDA approval and is currently on the market for processing of fresh juices.

13.4.3 Available PEF Units

Ohio State has designed three PEF units that differ primarily in their fluid handling capacity. The OSU-4, intended for laboratory use, has 1/8-in.-diameter tubing (about 3 mm). The OSU-5, designed for pilot-plant use, has 1-cm diameter tubing (10 mm); it is not commercially available, but OSU will build a laboratory unit for R&D purposes. The OSU-6, a 75-kW commercial unit built by DTI, has 1- to 1.2-cm-diameter tubing and can process orange juice at 500–2,000 L/h and apple juice at up to 5,000 L/h. DTI also offers several additional commercial units, as well as a smaller R&D unit and commercial-scale systems rated at up to 20,000 L/h. Current equipment costs are high, in part because of the relatively small market for the electrical equipment. Reliability can also be an issue with electrodes now needing replacement about every 100 h of operation. DTI, OSU, and others are investigating electrodes with longer lifetimes.

As fluid volumes increase, larger-diameter tubing implies larger gaps in the treatment chambers and thus higher voltages. A better approach to higher volume might be to use parallel small tubes. Finally, temperature control is still a significant issue.

13.4.4 PEF Being Applied Commercially

Genesis Juice Corp., Eugene, OR, was selling PEF-treated juices and blends in the Portland market in 2006. The products included apple, strawberry, and other flavors. The package was glass with a full label and they were sold from a refrigerated case. Genesis used an OSU-5 running at about 200 L/h. Prices were higher than for other juice products, but Genesis promoted them as being from organic fruit and having almost-fresh flavor. The major motivation for PEF processing is to avoid the loss of flavor from normal thermal pasteurization. Shelf life was said to be 4 weeks. Vegetable juices such as carrot need to be acidified and can pose challenges because of particles having to pass through the small clearances of the treatment chambers. *Listeria* in skim milk was among the more resistant organisms to PEF.

Clearly, PEF is not the answer for every food, but it is intriguing for certain applications. There remain significant engineering challenges.

13.4.5 Effects of MEF (a Version of PEF)

In the course of research on the mechanisms underlying other non-thermal processes, other effects on foods have been noted. In some cases, these can have interesting applications. For example, moderate alternating electric fields (MEF), even at field strengths three orders of magnitude lower than the 35,000–50,000 V/cm needed for pasteurization, can make fruit and vegetable cells more permeable, leading to higher yields of juice extraction. Application of MEF to fermentation broths can reduce the lag phase of a growth curve and stimulate production of bacteriosins. Some of these effects are attributed to the range of frequencies or harmonics that naturally occur with most electric power supplies. The power supplies needed to provide a relatively low-strength field ($\sim$10 V/cm) at frequencies associated with conventional electric power supplies are less expensive than the power supplies needed to generate a high-strength square wave, as is normally used in pulsed electric field processing. Thus, the equipment needed to realize some of the benefits of applying MEF to foods can be modest in cost.

13.5 Other Non-thermal Processes

Ohmic heating, in which electrical resistance is the source of heating, had been in decline after a flurry of interest but was showing some signs of commercial interest. Several units have been installed. Researchers at Ohio State have also investigated the effect of carbon dioxide on high-pressure treatment. It is known that carbon dioxide has some antimicrobial properties. It dissolves at high pressures and comes back out of solution when the pressure is released. This appears to enhance the destructive effect on spoilage organisms in juice.

13.6 Developments in Thermal Processing

While traditional thermal processing is usually associated with metal cans and glass jars, the newest developments involve new packaging materials and forms. One example is the flexible retortable pouch, originally developed for military rations and commercialized in the late 1970 s by Continental Baking and Kraft Foods. Except for military use, the retort pouch has languished until the early twenty-first century, when it has been applied to tuna, cooked chicken, and pet foods.

The pouch was touted as a low profile, inexpensive package, but it suffered from a requirement for 100% inspection of seals and relatively low-speed filling lines. Newer pouches have gussets to enable them to stand up and some have zippers to allow reclosing. The standup pouch can eliminate the need for a carton, saving costs and providing an opportunity for use of striking graphics on the pouch.

The original pouch structure involved a heat-sealable polyolefin inner layer (polyethylene or polypropylene), adhesive, aluminum foil, adhesive, and a polyester outer film. The aluminum provided physical strength and a barrier to oxygen and moisture. Newer pouches add a nylon layer for additional strength, especially in larger pouches and for the military, which still relies on foods in pouches for its Meal Ready-to-Eat (MRE).

Many of the products introduced in pouches are co-manufactured by firms whose major customer is the Pentagon. The feast-or-famine nature of military procurement cycles has made that business a difficult challenge. Civilian applications help absorb capacity that is required for defense and relief emergencies.

Another older technology that has proven durable and seems to be getting more attention is rotary retorting. Applied originally to large metal cans, such as No. 10 sizes for foodservice, rotation helps agitate the contents of a container during heating and cooling, thus shortening the processing time and reducing damage to flavor and texture. Several firms manufacture rotary retorts.

Finally, control systems for retorts are becoming more advanced, permitting real-time calculation of process lethality and optimization of temperature and time to achieve target processes without excessive heating. Traditionally, temperatures were recorded on circular charts, and any deviation from target temperatures and times could require reprocessing of the entire batch. Target temperatures were typically a few degrees higher than strictly required to allow for occasional dips due to changes in steam pressure. A mercury-in-glass thermometer was considered the most reliable temperature indicator. Temperature histories of test packs were used to establish a process schedule, which, once approved, became an absolute standard. Now, using inexpensive computers connected to each retort, real-time data can be used to calculate the accumulated heating and cooling effect, expressed as lethality or minutes at some reference temperature (250°F for low-acid foods). Processing can be terminated when the target is reached, and deviations are corrected on the fly.

Thermal processing is still the foundation of the processed foods industry, being applied to billions of containers of fruits, vegetables, soups, beverages, and meats every year. Despite their long history, new products are still being developed in these categories, especially with the US Department of Agriculture's approval of the new organic seal, creating an opportunity for organic soups and sauces. Ready-to-serve soups and specialty sauces benefit from new packaging and rotary retorting.

13.7 Examples and Exercises

As discussed, there are few commercial applications of the newer technologies, so exercises are based on hypothetical cases.

1. Calculate the theoretical output of a plant using any of the technologies described. How do you estimate cycle time? What supporting facilities are needed?
2. Do a literature search and describe the current state of one of these technologies.
3. What are some other relatively novel developments in food preservation?
4. Microwave heating has been suggested for aseptic processing. Discuss the advantages and disadvantages of this approach.

13.8 Lessons

1. Food preservation requires the exchange of energy in some form, and there are many mechanisms to achieve this.
2. It is important to understand the physics underlying the exchange of energy because often this is the limitation on rate.
3. The usual objective in food preservation is to eliminate or greatly reduce pathogens and spoilage microorganisms with the least damage to the flavor, color, nutrition, and texture of the food.
4. Sometimes the obstacles to commercialization of a technology are not technical, but rather emotional.
5. Some older technologies still have room for development of improvements.

Part III
A Few Broader Topics

Chapter 14
Economic Evaluation

This and the next chapter are based on material previously published in Clark (2009) and were specifically requested by reviewers of the original outline of the present book.

14.1 Measures of Worth

Cash is the lifeblood of a business, so a proposed investment is best evaluated by how it contributes to the flow of cash. Cash is defined in a business context as after tax profits plus depreciation. These terms will be discussed in more detail. The objective is not to duplicate more extensive treatments elsewhere, but rather to enable the engineer to understand and communicate with his colleagues in finance, accounting, and management (Clark 1997a, Valle-Riestra 1983).

A business such as a food company uses *fixed assets,* such as buildings and equipment, and less tangible assets, such as patents and trade secrets, to produce items for sale. The company pays for its assets with a mixture of *equity* and *debt,* where equity represents money invested or accumulated by the owners and debt represents money borrowed from a bank or other source. In accounting, debt and equity are considered liabilities, offsetting the assets that were purchased, as well as other assets such as money owed to the firm by customers, pre-paid expenses, work in process, and retained earnings.

In evaluating a potential investment, companies are interested in how the new asset will contribute to cash flow relative to its cost. There are also investments that must be made, which may not directly contribute to cash flow. These are considered non-discretionary and include projects dictated by customer safety, employee safety, and environmental regulations. Even these can be evaluated by the same techniques because almost always there are alternatives and the objective is to select that alternative which is most cost effective.

In the normal course of business, a company receives *income* from the sale of products and pays *expenses* associated with making the products and with operating the plant. Expenses may be *direct* or *variable* because they are proportional to the amount of product, or they may be *indirect* or *fixed* because they are relatively constant, no matter how much product is made.

J.P. Clark, *Case Studies in Food Engineering*, Food Engineering Series,
DOI 10.1007/978-1-4419-0420-1_14, © Springer Science+Business Media, LLC 2009

Examples of variable expenses include raw materials, packaging material, energy, and labor. Examples of fixed expenses include management salaries and benefits, marketing costs, taxes, insurance, and depreciation.

Profit before taxes is the difference between income and expenses. *Depreciation* is a special category of fixed expense that reduces taxes but is not a true cash flow. Rather, it is an allowance provided in the income tax laws that permits recovery of the cost of fixed assets over their assumed lifetime (Maroulis and Saravacos 2008).

Depreciation may be calculated in several ways: straight line, meaning an equal amount is taken each year; double declining balance, in which larger amounts are taken early in the project's life, for reasons explained shortly; and various modifications, in which, usually, larger amounts are taken early but not so large as by double declining balance.

The reason that the timing of depreciation is important is the *time value of money*. The time value of money is the concept that a dollar in hand today is worth more than one received a year from now. This is because money in hand can be used to earn more money in the form of interest or profit on an investment. Thus a dollar invested at interest rate, i, will be worth $(1 + i)$ at the end of 1 year. At the end of 2 years, it would be worth $(1 + i)(1 + i)$ or $(1 + i)^2$. It then follows that a dollar received 1 year in the future is only worth $1/(1 + i)$ today. More generally, an amount received n years in the future is worth $1/(1 + i)^n$ times the amount, F, today. The symbol, P, is often used for the present value.

Since higher depreciation charges early in a project's life contribute to higher cash flow in the early years, those values are discounted less in computing the *net present value (NPV)* of a project. The NPV is the sum over the life of a project of the discounted cash flows for each year.

An important factor in the calculation is the discount rate i. Considerable debate can ensue over the correct value to apply. The discount rate is meant to represent the potential return of relatively safe alternative investment opportunities, such as government bonds or the cost of borrowed money, which often is fairly close to the bond rate. In practice, the discount rate is adjusted upward to account for some degree of risk. The discount rate to use in a given company is usually set by the corporate finance department and may vary with time and for different classes of project.

It is important to realize that calculating NPV, and other measures of worth, using the time value of money and discount factors is meant for comparison of alternatives and should not be seen as computing an absolute measure of value. In many cases, cash flow contributions, positive or negative, which are identical among all options will be ignored in an analysis so that the focus is on those elements that differ. This means that the result is not a measure of absolute value, but only of relative value. In the case of a non-discretionary investment for which there are only costs and no positive cash flows, the result of a NPV calculation will be negative and the best choice is that with the value closest to zero.

When considering a company's capital budget as a whole and alternatives within it, there are almost always more costs of potential projects than there are funds to invest. Thus, projects are evaluated and ranked with the goal of investing the available funds as wisely as possible. In theory, those projects with a positive NPV using the agreed upon discount rate should be funded. In practice, those projects contributing the most to the company's value should be funded. This would tend to favor larger projects.

Even after applying all the appropriate quantitative measures, there remains an important role for strategic judgment in choosing between new products or markets as compared with extension of existing products. There is comfort in the familiar and risk with the new, but also greater potential from innovation. Publicly held companies are often pressured by stockholders to cut costs and increase profits, sometimes within a relatively short time frame, while an objective understanding of long term benefit might favor greater investment in projects with greater risk and greater potential.

There are other measures of worth besides NPV, which are in common use. One of the simpler ones is *return on investment (ROI)*. ROI is the ratio of profit to capital invested. It is usually computed for the first year and implies that profits are constant with time. It can be misleading, especially for projects with varying conditions and complex investment patterns, but is simple to calculate and easy to understand.

Closely related to ROI is *payback time*, typically the time to recover the investment from profits. It is the inverse of ROI. Neither of these approaches recognizes the time value of money nor the contribution of depreciation to cash flow. Some variations of ROI and payback time use cash flow instead of profits, which is more realistic, but not as conservative. Being overly conservative in evaluating investments can prevent the choice of wise, longer-term projects in favor of shorter-term and perceived to be safer choices. Strategic judgment should prevail.

Internal rate of return (IRR) is defined as the discount rate for which the NPV is zero. It is computed by a trial and error calculation of NPV for various discount rates. If IRR is greater than the value used by the company as a guide, then an investment is acceptable. Projects with high IRR should be funded. However, often relatively small projects can show high IRR while contributing relatively little value to the firm. Also, there are conditions of varying cash flow, such as additional investments over time, for which the calculation of IRR may give ambiguous results. While admittedly rare, the possibility of such a case suggests that IRR be used judiciously. NPV is generally recognized as unambiguous and reliable as a measure of worth for alternative investments.

14.2   Estimating Capital Investment at This Stage

To evaluate and compare investments requires an estimate of the cost. However, as pointed out in a series of studies by The RAND Corporation (Merrow et al. 1979, 1981, Merrow 1989) even when a project is 95% complete there can be uncertainty about its final cost. How, then, does one make a reasonable estimate early in a project? There are several possibilities:

- Past experience
- Comparable facilities

- Feasibility study
- Detailed design

A reliable approach is to use past experience, especially with similar facilities. Some companies build essentially the same plant, often using the same designers and contractors, year after year to satisfy steady growth. Even with this much experience, there are differences among the plants because of site differences, lessons learned in the past, and changing expectations of flexibility.

Site influences include seismic zones (relative risk of earthquakes), environmental regulations, flood plains, and availability of utilities. Each of these can affect the cost of what otherwise might be an identical facility. For example, structural requirements are quite different in California than in the Eastern United States because California experiences frequent earthquakes. Buildings are required to take the anticipated forces of an earthquake into account in California. In practice, this may mean reinforcing concrete block walls, for instance, and fastening equipment to the floor, whereas in other states the equipment might simply be put into place.

Regulatory restrictions on air, odor, and water emissions also vary from place to place. For example, areas subject to smog formation may have more stringent restrictions on emissions of volatile organic carbon (VOC). VOCs are compounds that can interact chemically in the atmosphere to form irritating compounds. Bakeries emit small amounts of ethanol formed during the fermentation of bread and in many parts of the country must reduce the amount with fairly costly equipment. Choices include incineration with natural gas, catalytic oxidation over a supported precious metal, and cryogenic condensation. Water and odor restrictions also vary, depending on local conditions, waste treatment capability, and the proximity of neighbors. Even the sources of benign odors, such as coffee roasting, chocolate, and bread baking, may be considered noxious by some neighbors and require controls.

Finally, there is a significant cost to bringing fuel, electric power, water, sewer, and rail to an undeveloped site as compared with one where these utilities are present. Of course, there will usually be a difference in the land cost as well between a developed site and raw land. The RAND studies found that one of the biggest contributors to inaccurate cost estimates was not having a specific site in mind.

When cost estimates are performed for a hypothetical site, it is almost always assumed to be flat, requiring little grading and having all utilities nearby. In reality, there are few, if any, such sites to be found. The good ones have been developed; leaving land that almost always has some costly defect, such as poor soils, poor drainage, an environmental hazard that must be removed or sequestered, or an unusual shape. Without a specific site in mind, it is difficult to anticipate what costs might be encountered. It seems to be human nature to estimate optimistically, but this is almost always wrong.

Lacking directly relevant past experience, comparison with other facilities can be useful. Bartholomai (1987) published a useful collection of mostly small plant designs for a variety of food factories and Maroulis and Saravacos (2008) derived some correlations from his data to show relationships of capital costs to projected sales and of total costs to equipment costs. Sometimes, equipment costs are relatively easy to estimate, compared to those of a total project because the equipment is directly related to the process, while the project has a myriad of other costs and influences.

Food plants can cost about $100/ft^2(Clark 1997a) for the building and support utilities, but these costs can vary dramatically depending on details of the design. Some food plants require refrigeration of the entire space, such as meat plants, and these then have greater insulation requirements. Other plants must be capable of resisting large amounts of water and harsh chemicals, such as dairies. Such plants demand more expensive finishes for floors and walls. Broadly, food facilities can be categorized as dry or wet, refrigerated or not, dusty or not, and handling raw agricultural materials or not. (Raw agricultural materials may include fruits, vegetables, root crops, and grains and handling them implies dealing with dirt, insects, and other potential contaminants.) Relying on comparable facilities for guidance in cost estimating means understanding the impact of these differences on costs.

If time and budget permit, the most accurate cost estimate relies on a relatively complete and detailed design for a specific site. However, neither cost nor time usually permits this approach and instead a feasibility study is performed. A feasibility study is a limited exercise usually performed by a small team in which sites may be evaluated, a preliminary design developed, and a cost estimate prepared. It is understood that the design is not final but rather is intended to support a cost estimate sufficient to allow evaluation of the project, hence the feasibility or viability.

One measure of the relative accuracy of cost estimates is the *contingency* that is included. Contingency can be a controversial topic in the same way that discount rate often is, because both involve judgment. Contingency in a cost estimate is a line item that is meant to account for costs that are not otherwise explicitly identified and for errors in the other items. Contingency is not intended to account for changes in scope, acts of God, or other catastrophes. As designs progress, details are resolved and cost estimates become more detailed, contingency as a percentage of total cost can be reduced, but experience dictates that it never disappears. In a feasibility study, contingency might easily be 30% of known costs or even higher. In an estimate derived from a final design, contingency should still be about 10%. Because of a misunderstanding about what contingency is intended to cover, some companies insist that contingency cannot be greater than 10%. This corresponds to the margin of error that many firms allow in project budgets. In preparing a feasibility study estimate, it may be necessary to bury the true contingency to satisfy this type of arbitrary dictate. The point is to understand the limitations of early estimates and to include proper allowances for inevitable errors.

Finally, if a detailed design, or at least a more detailed exercise than a typical study, can be performed, the resulting estimate will be more accurate and can have a lower contingency. Underlying any estimate must be a clear narrative scope document. The feasibility, preliminary, and detailed designs make no sense without the scope document. A scope document is meant to be capable of change, as circumstances change, but without the original as a base, the reality and consequences of change are hard to identify.

In summary, a food facility cannot be designed and evaluated without precisely articulating what it is, where it is, and what it is to do.

14.3 Estimating Costs and Benefits

Sometimes it is difficult to estimate operating benefits, which is one reason much emphasis is often placed on costs and cost reduction – costs seem easier to quantify. Finance and accounting people are often skeptical about claims of benefits because they may seem like wishful thinking. It is important in estimating benefits to be realistic and to base estimates on historical data whenever possible. For example, a real benefit of a new or expanded facility may be increased sales and thus increased profits. If sales of the firm or of the relevant product line have been steadily growing and capacity of existing manufacturing has been reached or exceeded, then it can be persuasive to project future sales continuing to grow, probably at a somewhat lower rate. The lower rate is suggested because in reality most products tend to grow at a reduced rate as they mature.

If the facility is justified by projected sales of a new product with no prior history, it is important to have some test market data, which shows likely per capita consumption, or some other basis for estimating sales. Justifications based on improved *quality* are difficult to support, though they may in fact exist. Often in the course of design, choices arise that can affect capital or operating costs and which are evaluated by their impact on quality of the product. It may be easier to justify an incremental increase in capital cost by a *life-cycle analysis* than by an assertion of improved quality, simply because it is hard to persuade people that quality increases sales or profits.

Life-cycle analysis refers to the practice of estimating future maintenance and operating costs, discounted to the present, as well as initial cost, in comparing alternatives. A piece of equipment or a building finish with a higher initial cost may be less expensive to maintain, be more energy efficient, and be less expensive to operate over its lifetime and thus justify its higher initial cost. Examples include heating, ventilation, and air conditioning (HVAC) equipment, boilers, floor and wall finishes, and increased roof and wall insulation.

In developing countries, initial cost can be most important because capital is scarce, risks are high, and the odds of a given project succeeding may be low. Thus, while a life cycle analysis might justify a high initial cost for a given item or entire project, a lower cost alternative may be preferred and, in fact, be the wiser choice.

Another benefit from a new or expanded facility might be a reduction in the cost paid to produce a product through a contract manufacturer or co-packer. *Co-manufacturing* is a common practice and there are many companies and facilities dedicated to this service. Using a co-manufacturer permits a company to have fewer fixed assets than it might otherwise have, often allows quick entry to a market, and may give a company access to technology and expertise it does not have. Companies that specialize in manufacturing for other companies believe that they have superior management and operation techniques because they are focused on manufacturing, leaving marketing and sales to their clients.

On the other hand, there are some disadvantages as well. The most obvious is that the co-manufacturer expects to make a profit, that is, to earn some return on the assets that are employed. If a company were to invest in its own assets for manufacturing, it would expect to earn a return on that investment. In fact, it might expect a higher return than does a co-manufacturer. Further, the book value on which a return must be earned may be lower for the existing assets of a co-manufacturer than it would be for a new facility. Many co-manufacturers are privately held and are willing to accept a lower profit margin than are publicly held large corporations. Nonetheless, many users of co-manufacturers begrudge them their fees and look for ways to avoid or reduce them.

There can be less direct control over quality and practices at a co-manufacturer, since the employees are not those of the customer. Understanding this, co-manufacturers encourage close interaction with their customers and often have quality control personnel on site from customers. Finally, there is some risk to proprietary information that may be shared between the co-manufacturer and the customer. This last risk is more perceived than real because co-manufacturers who violate confidences will not remain in business very long. On the other hand, co-manufacturers cannot help learning from each assignment and may be tempted to become competitors at some point.

Co-manufacturing makes the most sense when appropriate assets exist and are not fully utilized. It is common that some investment is required, such as for special packaging, but if an entire new line is required, then a different justification is needed. If major investment is required, the customer will have to pay a sufficient fee to provide a return to the co-manufacturer. However, it can happen that the co-manufacturer does not demand as high a rate of return as does the customer and, therefore, using the co-manufacturer even with major capital required, makes sense. Also, many co-manufacturers have lower labor costs than their customers and, because of their experience, may be more efficient at operations than their customers are. Co-manufacturers may be especially appropriate when entering a new area of technology. Finally, using a co-manufacturer may significantly shorten time to market.

For these reasons, many major food companies use co-manufacturers for some, if not all, of their requirements. Likewise, there are many successful co-manufacturing companies, who are typically almost unknown to the public. If part of the justification for a new or expanded facility is cost savings from bringing a product or products in-house, then all the costs should be considered, not just the toll or fee charged by the co-manufacturer.

Another benefit from a new facility may be improved efficiency due to new technology, reduced cost of energy or raw material, or reduced labor cost. At one time, many manufacturers moved from the Northern United States to the South because labor was less expensive in the South. Areas with less active unions are perceived to have less expensive labor. New facilities may have greater automation and thus reduce labor costs. Areas in which electric power is generated with dams may have lower energy costs. Countries outside the United States tend to have less expensive sugar and so some confectionery companies have relocated from the United States. Other costs, such as shipping, may offset some of the expected savings from such a move.

Some of the operating costs are as follows:

- Raw materials
- Energy
- Labor
- Waste disposal
- Packaging material
- Depreciation

Raw materials and packaging material constitute, on average, about 70% of the cost of foods (Clark 1997a). Other costs, including labor, energy, and depreciation are about 10% each. A major influence on raw material costs is yield, that is, how much or how little of what is purchased winds up in the finished product. Some yields can be shockingly low. For example, it takes about 10 lb of milk to make 1 lb of cheese. About 50% of a hog or steer is edible. Other raw materials, such as grains and vegetables, can have similar losses that are essentially intrinsic, that is, they occur because of the composition of the material.

There are other sources of yield reductions, in addition to composition of the raw material, some of which can be addressed. Examples include the following:

- Spillage
- Lost production due to off-specification products on start-up
- Over-filled containers (giveaway)
- Improper recipes
- Mislabeled containers
- Improper processing
- Contamination
- Out of code (aged) raw material or product

Good engineering, construction and operation of material handling equipment, such as conveyors, can reduce spillage. Every transfer point is a candidate for leakage and a spill and therefore a candidate for careful observation and improvement. Material on the floor of a food plant is not only a sign of poor housekeeping, but also of poor maintenance and operation. A useful approach is to have trays or shelves under transfer points, which can catch spills before they reach the floor and possibly permit the material to be added back to the conveyed stream.

In almost all processes, there is a departure from normal operation when the process stops, some correction or repair is made, and the process starts up again. Some products age poorly and may have to be discarded, some partially filled containers may be created, and some deviation from target conditions may occur, for various reasons. In general, process interruptions are undesirable. One measure is mean time between failure (MTBF), which is computed by counting interruptions over a shift and dividing this number into the length of a shift. The mean time to repair (MTTR) is the sum of total downtime divided by the number of stops. The objective is to have a high MTBF. It is harder to generalize about MTTR. Short MTTR could indicate relatively insignificant interruptions, but if they are frequent (short MTBF), then it could indicate an unstable operation. A stable, reliable process probably has both a high MTBF and a relatively high MTTR, suggesting that it rarely stops and then only for something serious that takes a while to fix. Preventive maintenance is one tool to reduce stops and to make them less severe when they do occur.

Reducing product giveaway is the motivation behind digital scales, which are so accurate and precise that they have been justified by the saving of one potato chip per package. The challenge of product dispensing and filling is that under filling is considered fraud, so all manufacturers over-fill to some extent. Filling is a statistical process in the sense that there is a normal distribution around a target fill weight with most filling equipment, in part because of the natural variability of foods.

Two different normal distributions might result from two fillers, one with a small standard deviation and another with a larger standard deviation. Both could be centered on the same mean–225 g, for instance. Depending on the variability of weights achieved for a given food in a given filler, the target weight is set so that a high fraction of all packages exceed the minimum allowed, typically the label weight less some small deviation, often set by regulation. This means, in practice, three standard deviations greater than the minimum. The smaller the standard deviation, the more narrow the distribution and therefore the smaller the excess fill over the minimum. Digital scales use up to 16 buckets into which product is dispersed and then, very quickly, each bucket is weighed and some combination of four buckets is chosen and their false bottoms released to fill a container so that the contents are close to the target. For example, the filler with more precision (small standard deviation) could be set to deliver about 232 g on an average while the less precise filler would be set at about 245 g. Both fillers would over fill on average but the more precise filler would save 15 g/package compared to the filler with a large standard deviation. In practice, data are taken on a large number of packages, using an in line checkweigher, to establish the distribution and the filler is fine tuned to meet the target with minimal give away.

Liquid fillers often use precise volumetric pumps or valves to achieve the same goal of small standard deviation and reduced give-away. Pumps and valves are made with close tolerances, but suffer wear with use, so their precision may decline over time. Checkweighers are used to monitor package weight and to indicate when maintenance may be required on a filler. Under weight packages are rejected after the checkweigher. Under weight packages may be opened and the contents salvaged; they may be sold in a company store to employees, donated to a food bank, or discarded, depending on the value of the contents, the amount of packages involved, and the availability of outlets. A package may be under filled because of temporary failure of a bucket, valve, or pump; because of an empty feed hopper; or because of a variation in product density. Check weight data should be carefully monitored to identify trends that may indicate a process flaw or a need for maintenance.

Recipes may have too much or too little of an ingredient due to human error or mechanical failure of a feeder or scale. Too much use means a yield reduction, at least, and could impact quality. Too little used could impact quality such that a batch is discarded or needs to be redone. A major goal of operation management and automation is to reduce errors in formulation so as to maintain consistent quality and maximize yield.

Food labels are governed by regulations intended to inform consumers and to prevent fraud. Labels may be incorrect because they do not declare an ingredient that is used, deliberately or inadvertently; because the weight is incorrect; because the nutritional information is incorrect; or because the food may contain an undeclared allergen. The contents may be otherwise safe and nutritious, but the product cannot be sold. Mislabeled product may be repackaged, discarded, or given away out of commercial channels, such as to a food bank. In any case, it is a source of yield loss or cost increase.

Improper processing may result in overcooking or undercooking, poor color, poor shape, poor particle size, poor distribution of a mix, or some other failure attribute. A company establishes quality standards for such attributes and rejects those that exceed the boundaries. Some improper processing can be corrected by re-processing, but at other times the product is discarded. Out-of-specification food is often diverted to animal feed, especially grain-based products such as cereals and pet foods.

Some processes, such as extrusion of snacks, cereals, and pet foods, have start-up and shutdown episodes during which out-of-spec material is inevitably produced. Keeping such times as brief and infrequent as possible is a good way to improve yields.

There are many potential sources of contamination of foods. Some are easier to detect than others. Metal and other foreign objects can be detected by X-rays or potentiometric coils like those found at airport security. Metal detectors only detect metal pieces above a certain size while X-rays can also detect glass, wood, and plastic, provided the materials are sufficiently different in density from the food. Food may be examined on its way to packaging or after filling. Conventional metal detectors cannot examine metal containers or those laminated with aluminum foil. In fact, metal detectors are used to confirm that small foil pouches of flavor are included in packages of instant noodles.

Packages or streams of food found to contain metal or other foreign objects are diverted for further examination, reprocessing, or discarding. Depending on the value of the food, it might be passed over magnets or screens to remove ferrous metal or other foreign particles. It is important, when foreign object contamination is found, to establish the source and cause if possible.

Metal fragments enter food due to wear of equipment parts, by breaking of screens, by careless maintenance, and by deliberate or accidental introduction by people. Prevention of such introduction is one reason visitors and workers are asked to remove jewelry and objects in shirt pockets. Typical workers' uniforms do not have shirt pockets or metal snap closures to help prevent accidental introduction of foreign matter to food.

Other contamination can occur from dripping water, dust, grease, or incorrect ingredients. Such contamination may be difficult to detect, so prevention rather than re-processing is the best protection. The principles of sanitary design are primarily intended to prevent contamination. Places where food is exposed pose the greatest risk, so, often, conveyors may be covered, kitchens and preparation areas may have ceilings, hoods for dust collection are installed at ingredient dumping locations, only food grade lubricants are used, cold pipes are insulated to prevent condensation, and seals are routinely inspected for leaks of lubricants.

Finally, ingredients that have exceeded their usable life must be discarded and represent a yield loss. Inventory control with first in first out practices are important in using ingredients before they become too old. Foods are perishable to varying degree – some absorb water and cake up, some lose water, some oxidize, and some discolor. Proper storage conditions are important as well. The ingredient storage area should be dry, temperature controlled if needed, and sufficient space provided for access to stored materials. Purchasing ingredients in the right amounts can prevent loss through aging. The concept is to understand the frequency and amount of use and to purchase accordingly. This may mean paying a slightly higher unit cost for an ingredient than if larger quantities were procured, but by preventing loss through aging, the net cost is reduced.

Energy is a relatively small operating cost for food plants, but controlling it in times of rapidly rising prices is important. Good insulation, efficient refrigeration plants, efficient boilers, and intelligent energy integration are all contributors to reducing energy costs. Most food plants burn natural gas because it is clean and relatively plentiful. Some food processes, such as baking, are energy intensive, in the sense that they use a lot of energy, but they may also be energy efficient because baking ovens are typically well insulated. The highest use of fuel in a bakery is often the boiler, used primarily to make hot water for cleaning. Lighting is an opportunity to use energy efficient fixtures, typically halogen lamps. Mercury vapor lamps are avoided in a food plant because they attract insects.

Food plants vary widely in their dependence on labor. Meat processing is very labor intensive, while some other plants operate with very few people. It is sufficient here to say that people are extremely important in a food plant and that, even though they may be a small cost, they can make or break an operation. In particular, food processing still relies heavily on the skill and craft of its workers because the product is ultimately judged by its taste and other sensory properties. Modern practice is to hold line operators responsible for the quality of their production by training them in quality standards and then requiring them to evaluate products on a regular schedule.

Wastes include water and solid wastes as well as air emissions. Food plants can use large volumes of water, especially for cleaning. Wastewater discharges are treatable but can be relatively strong and may have a wide range of pH and fats, oil, and grease (FOG). FOG is worrisome because it can solidify in sewer lines. It is good design practice to have separate sanitary and process waste lines, which are connected to a sewer through an air break, typically in a catch basin. The air break is to prevent back-flow of sewage into the process area, and is achieved by having the waste lines from the plant enter the catch basin near the top, while the sewer connection is near the bottom. A catch basin is a cylindrical vessel built beneath grade level and accessible through a man way. Volume of the vessel is based on achieving a minimum residence time for the heaviest flow rate, typically seen during cleaning. Residence time with the vessel half full depends on the amount and properties of waste, but should be several minutes. Heavy solids drop to the bottom, below the sewer connection. FOG floats to the top. FOG and heavy solids are periodically removed by an outside service and disposed of as solid waste. If the removal of FOG is not sufficient to meet discharge requirements, then it may be necessary to install an air entrainment separator or dissolved air flotation unit, in which small air bubbles help float FOG to the surface, where it can be skimmed and trucked away.

Most food plants discharge their liquid wastes to a municipal treatment plant for a fee. It is important to confirm that the municipal plant has adequate capacity and that the fees are fair. Even when the treatment plant is adequate, it is wise to practice water conservation in the plant. Typically, the plant must buy water and also pay for waste treatment. If water is a significant part of the product, it is wise to measure and document exact usage because often water consumption and waste discharge are assumed to be closely related. For example, a soft drink or fruit juice plant may purchase much more water than it discharges as waste because so much of what it buys is used in the product.

Clean-in-place (CIP) systems in which water and cleaning solutions are recovered and reused not only reduce labor but also save water and reduce discharges (see Seiberling 1997). Other water conservation measures include having "dead man" nozzles on hoses so that they only run when attended and selecting equipment with low water use. Using a cooling tower to recycle and reuse cooling water is another good practice. A cooling tower is a device that reduces the temperature of a water stream by evaporating a portion through contact with air. The air may be circulated by fans or it may be moved by natural forces – wind and natural convection as the air temperature changes. A cooling tower is constructed of corrosion resistant materials and sometimes uses chemicals to reduce scale and corrosion. Fresh water is added to compensate for that lost by evaporation.

Solid food wastes are often taken by farmers to use for animal feed. If so, they should be stored carefully and protected from contamination. Wet solids may spoil quickly and so should be removed often. It is rarely economic to dry such wastes, but sometimes this is done. Examples include spent grains from brewing; start-up and shutdown residues from cereal, snack, and pet food extruders; and citrus peel from juice plants.

In isolated areas, food plants may need to treat their own wastes and provide their own potable water. A common approach, where land is available, is to spray the wastewater on land planted in grass or other crops. Soil permeation properties, wind, and temperature all affect whether land application is appropriate. Land application is not practical in freezing weather, but is suitable for seasonal canneries, which typically are mostly idle in the winter. It is also important to understand the source of incoming water and to be sure it does not become contaminated by the land application. Incoming water is often raised from wells whose aquifer could become contaminated.

Water for use in foods must be potable, meaning it complies with minimum standards for safe drinking. This may not be adequate for a particular process. For example, drinking water is often chlorinated, but for use in beverages, it may be treated with activated carbon to remove the chlorine taste.

Finally, depreciation has been previously discussed. It is listed here as a cost to remind that the capital investment matters. Depreciation is one component of fixed cost or overhead, as distinct from variable costs, such as labor, energy, and raw material. Typically, unit fixed costs are reduced as capacity utilization increases, which is one reason more food plants consider 7-day operation and multiple shifts. The same facility can produce more, so long as there is a market, and the fixed cost component of each unit produced goes down, because there is little need for more management, and depreciation remains constant, as do other components of fixed cost, such as taxes and insurance.

In evaluating a project, all the anticipated benefits and costs must be estimated. This is only reliable for a specific site because utility, labor, energy, and raw material costs all vary with location. It is important in estimating benefits to be creative without exaggerating and losing credibility.

14.4 Discussion Topics or Assignments

1. Obtain the general parameters for a relatively new food plant from such sources as trade publications, *The Wall Street Journal*, and the Internet. Calculate all the measures of worth that you can, stating any assumptions you find you must make.
2. Compare three projects on such parameters as cost per square foot, cost per unit of production, and cost per dollar of sales. Discuss the differences and similarities as related to product category, scale, and any other characteristics you can identify.
3. Research the unit costs of utilities in your area (community or county). Compare to national averages. Is yours an attractive area for a food plant? Why or why not?

Chapter 15
Design of a New Facility

The design and construction of a new facility is a relatively rare event in the career of most food professionals. A major expansion is much like a new plant except that the site and some supporting facilities may already be in place. This chapter discusses some of the issues that arise in a typical new plant project. A food professional may be a member of a team responsible for selecting and supervising various consultants involved or he or she may be such a consultant, drawing on education and experience. For the most part, there has been little published in this area (Clark 1993a,b,c, 2005a,b, Lopez-Gomez and Barbosa-Canovas 2005, Maroulis and Saravacos 2003).

15.1 Site Selection

For a new facility, one of the first issues is where it is to be located. This is the art and science of site selection. Site selection starts broadly, identifying a country or region and then focuses more narrowly until a specific piece of property is identified. One can think of choosing a time zone, then a state, then a zip code, and finally an address. What are some of the influences and how does one make choices?

One of the more significant influences is location of a firm's existing and potential markets. Some companies are initially regional and aspire to a more national presence. Others already have a national presence in the United States and seek international markets. Supporting existing markets is a distribution system with distribution centers (DC), customers, and delivery systems (trucks, rail, sales and delivery people). Moving into a new market requires integration with the existing distribution system and may require additional facilities, such as warehouses and depots. Adequacy of transportation systems can be a major factor because raw materials and packaging materials must be obtained and finished products distributed by ship, rail, or truck. As will be discussed later, some plants are near ports while others need not be.

If a firm is starting new, building a facility may not be the wisest strategy until its product line is proven. Typically a new venture arranges for production by a third party or uses an existing facility while getting its products established. Sometimes this arrangement may be maintained for many years. Usually, when considering a

J.P. Clark, *Case Studies in Food Engineering*, Food Engineering Series,
DOI 10.1007/978-1-4419-0420-1_15, © Springer Science+Business Media, LLC 2009

new facility, it is because a product line is well established and alternative sources of production are inadequate or may be considered too expensive. This usually means that the distribution system is well established and can be extended to a new region as needed.

In the United States, the Midwest is a popular location for food manufacturing because that region is a source of many raw materials and it has an extensive transportation system of rivers, rail, and highways. Other popular locations include the Northeast because of proximity to large population centers and California because it is a large market in its own right, grows many raw materials, and is expensive to supply from other regions of the United States because it is a long way from the Midwest and is separated from the rest of the country by mountains.

Raw materials, relative product density, and product shelf life are additional influences on location. Raw materials may be agricultural products, live animals, partially processed materials (such as coffee beans, meat, or raw sugar), or ingredients such as syrups or flavors. Water is often an ingredient in foods, so a plentiful source of potable water is always a consideration.

If raw materials are perishable, bulky, or seasonal, it often makes sense to locate processing plants close to the source. Thus tomato processing plants are often very close to the farms where tomatoes are grown. The products, tomato sauce, tomato paste, ketchup, and canned tomatoes, being shelf stable and concentrated compared to the fresh fruit, can be stored and shipped. Meat processing used to be concentrated near rail yards in Chicago and Kansas City so animals could be received and meat shipped in refrigerated rail cars. In recent years, meat processing has moved closer to where cattle and hogs are fed, in the mountain west for cattle and in the corn belt for hogs. Wineries are almost always placed near vineyards, though some in colder states, where grapes are difficult to grow, buy grapes from warmer areas. Frozen and dehydrated potato plants are usually located near potato growing areas, most in the northwest, while potato chip plants are located near markets because their products have a low bulk density and relatively short shelf life.

Bread is typically baked close to markets for much the same reasons – low density and short shelf life. These characteristics affect the cost of shipping and over how far a distance it makes sense to ship. Processed foods with longer shelf lives and higher bulk density can be manufactured almost anywhere because it is reasonable to ship the products long distances. Site selection for such products is dictated by the overall distribution system and strategic considerations.

Once a broad region is selected, then specific states and communities can be considered. There are a number of factors that can apply:

- Taxes
- Labor supply
- Energy supply and cost
- Water supply and cost
- Solid and liquid waste disposal capacity and cost
- Incentives
- Transportation (roads, rail, ports)

Taxes vary widely across the United States and across the globe. In addition to federal income taxes, there are state and local taxes on income, personal property, and real estate. Some states charge a tax on inventory of finished goods as of a certain date, which may make those states less attractive than their neighbors who may not have such a policy. Companies often reduce inventory near such dates to reduce their tax burden. Real estate and employment taxes are often used by local governments to support schools and other services. Because they are imposed by local governments, they may be negotiable to an extent. Abatements or freezes on assessments are often offered as inducements for plants to locate in certain communities.

Labor supply and cost matters more to labor intensive industries such as meat-packing than it does to a more automated factory. However, all food plants require some labor, and the required skill level is increasing, as plants are more automated and equipment is more sophisticated. Immigrants are often the source of large labor pools, and so there are pockets of Southeast Asians and Hispanics in the Mid-west working in meat plants and other food processors. Labor costs in the South were traditionally lower than those in the Northeast, leading to a migration of such industries as textiles from New England to the South. Food manufacturing is typically less sensitive to labor costs than to other costs, such as raw materials, so relocation just to reduce labor cost is less frequent in the food industry. However, in selecting a specific state or community, the availability of sufficient labor is a concern.

Since management of a plant may need to relocate from other sites or be hired specifically for the plant, the desirability of the location as a place to live is also a serious factor. Thus the quality of schools, amenities such as recreation and culture, and relative levels of education will be considered.

Most food plants are not energy intensive, but some can consume large amounts of fuel or electricity, especially if they are refrigerated, have ovens, or have large amounts of frozen storage. Energy costs have been increasing and are quite variable across the United States, especially for electricity, depending on whether an area has access to hydroelectric power (the least expensive) or must rely on natural gas, coal, or nuclear. Large users of power can usually negotiate favorable rates, but may be required to accept interruptions in supply, which may or may not be tolerable to a food plant. There are energy saving and reduction technologies, which help reduce a plant's energy footprint, but controlling the cost of energy is always a consideration. Co-generation, in which both heat and electric power are generated, typically with a gas turbine, can be attractive when there is sufficient demand for the heat. Some states allow private generators of electricity to sell excess power to local utilities. This probably will become more common as the federal government tries to reduce US dependence on imported oil. Some utilities discourage co-generation by paying relatively low rates for the power produced, but the mere possibility of building a co-generation facility can sometimes result in lower electricity rates from a utility.

Food plants can use large amounts of water for cooling, cleaning, and as an ingredient. Water in a food plant must be potable, meaning suitable for drinking safely, but even when it is, it may be more or less suitable for use in foods. Beverage plants, for instance, treat incoming water to remove turbidity (suspended matter) and chlorine. Locally available water may have excessive hardness or unsuitable flavors, even if it is suitable for drinking. If water is relatively expensive, it may make sense to have a closed cooling water system to avoid discharging water that is merely heated up. In a closed system, warm cooling water is exposed to air in a cooling tower, where some evaporates and reduces the temperature of the bulk of the water. A small amount of fresh water is added to make up for losses by evaporation. Chemicals may be added to reduce scale formation and to minimize corrosion of the tower and piping.

Solid and liquid waste disposal can be an issue. Food plants can produce large amounts of both liquid and solid waste. While it is biodegradable because of its food origins, it may also be vulnerable to rapid decomposition, producing odors and attracting insects, birds, and rodents. The recommended approach to liquid waste is to discharge to a municipal treatment plant for a fee. Where there is no adequate plant, the food plant may need to treat its own wastes. This can require a substantial investment and an ongoing operating expense. Food plants in rural areas, such as fruit and vegetable canneries, have often used land disposal to treat liquid wastes by spraying on fields planted in grass or other forage crops. This has become less common as housing has encroached on these once-isolated plants.

Land application depends on permeable soil, predictable winds, and mostly mild climates. Because many canneries did not operate beyond the harvest, the impact of winter weather was not important.

Modern food plants required to treat their own liquid wastes may use conventional activated sludge processes or industrial waste treatment processes, such as anaerobic digestion. Activated sludge uses high concentrations of aerobic bacteria to consume the organic matter in a waste stream. Excess microbial mass is separated by gravity and disposed of by spreading on land or by anaerobic digestion. Sludge is bulky and may have high concentrations of heavy metals and other toxic substances that are not consumed by bacterial growth, making its safe disposal a challenge. Anaerobic digestion uses a different family of bacteria, in the absence of air, to convert organic matter in a waste stream or in sludge to carbon dioxide and methane. This gas may be recovered for its energy value or burned for disposal. Anaerobic digestion is slower than the aerobic process of activated sludge but produces the useful byproduct gas and relatively little undigested sludge.

Often, a plant's sewage charges are based on incoming water consumption. If the plant puts a large amount of water in its product, it is wise to meter sewage separately, so as not to be charged for waste disposal that it does not use. Soft drinks, beer, and reconstituted juice plants are such cases.

Solid wastes may be taken away by farmers to use as animal feed or they may be dried to stabilize them before removal. The only other alternative is landfill, which may be expensive and limited in capacity. Packaging material wastes, such as empty bags and trim from packaging that is made on site, may be recycled but are often burned or land filled. A good food plant design aims at reduction of liquid and solid wastes, but there always are some, and the costs and capability of disposal of them can affect site selection.

Transportation systems can influence the size and location of a plant, especially in some developing areas, where roads are so poor that distribution areas are restricted and so plants must be smaller than they would be if located in the United States, with its interstate highways.

Finally, states and communities may offer incentives beyond tax concessions to induce a plant to locate in their area. The motivation is increase in jobs and, eventually, in the tax base. Incentives may include the following:

- Free or low cost land
- Free or subsidized training for workers
- Frozen tax assessments
- Reduced tax rate for specified period
- Cooperation in securing zoning and permit variances, if needed
- Assistance in relocating personnel
- Infrastructure improvements, such as paving a road
- Concessions from utilities on rates or upgrading supply
- Others specific to the case

Negotiating incentives is an art practiced by specialized consultants. Many areas have development groups for regions, states, and towns, which publicize the advantages and benefits of their area. They can be sources of incentive opportunities, but ultimately it is the political leadership that determines how aggressive an area may be in attracting new investment. If all else is equal among several candidate areas, a firm considering a new location might do well by letting it be known that several candidate areas are being evaluated. This promotes competition and may improve the eventual incentive offers. Incentives should not tip the scales when other factors are not equally attractive because usually the incentives are temporary, while other factors may be permanent in their effect.

In considering a specific piece of property, once a region, state, and community have been chosen, the factors are practical and mundane:

- Cost
- Size
- Soils
- Neighbors
- Utilities
- Location

Within a given area, cost of raw land is not likely to vary widely, but the more desirable the location and the more developed it is, the more expensive it will be. Going a few miles out of town can reduce the cost, all else being equal, quite a bit. Size is obviously critical – the land should be about four times the footprint of the plant to allow for truck circulation, parking, set back from roads, and some expansion. If major expansion is contemplated, then the amount of land should be increased for a given original plant size. It is generally better to buy sufficient land in the first place than to depend upon getting more later on if it is needed. The price will certainly increase over time and that vacant parcel may not be available when the need arises later. If land disposal of liquid waste is planned, then even more area is required. Some large companies acquire land in various areas well in advance of a specific need to provide choices when a new facility is planned.

Building cost is influenced by the structural strength of the soil and the terrain. Level land with good soil is in short supply in the United States because many of the desirable plots have already been built upon. This means that one can safely assume some form of soil issue with almost any candidate property. Some will require grading; some will require removal of past contamination (brownfields); and some will require removal of soil and replacement with engineered fill because the soil is weak or wet. Most food plants are built with slabs on grade, so the load bearing strength of the soil is critical. If it is inadequate, there may be a need for driving piles to support the building or the poor soil may be removed and replaced with fill that is appropriate. Brownfields are sites that have experienced some environmental contamination, such as leaking chemicals, buried storage tanks for gasoline, or past use as landfills. Depending on the exact conditions, the contamination may be removed or enclosed to prevent further spread. It is wise to inspect a property during or after a rain to observe drainage and detect potential flooding issues.

Food plants are not usually bad neighbors, but some do emit odors, noise, or dust, so it is best if residential areas are distant from the plant site. It is also important not to have bad neighbors, who might be a source of contamination or other hazards, such as a chemical plant or landfill. Where a food plant might become a nuisance, costs must be budgeted to reduce the impact, perhaps by landscaping, baffling for noise reduction, or other controls, such as incineration of stack discharges.

The degree of development can affect cost of the property and of the plant. It is most convenient if major utilities are at the property line, such as high voltage power, water, natural gas, and sewer. If these must be brought any significant distance, the respective utilities will charge their costs unless they can be persuaded to abate those costs as an inducement. Likewise, it is best if a road provides access to the property, but this may not be the case where a large plot is subdivided, in which case potential roadways are usually planned but not necessarily paved. Usually the seller or developer will provide access roads, but if this has not yet been done, there may be a delay in the start of construction.

Real estate professionals emphasize location in determining value of land, and it certainly has an impact. A good location gives convenient access to transportation routes, maybe has a rail spur if needed, is relatively easy for workers to reach, and, as mentioned previously, has the services and characteristics desired. Public transportation is not common in rural areas, so an employer may need to assist workers in getting to work by subsidizing buses or even, in other countries, by providing dormitories and cafeterias.

15.1.1 Example of Site Selection

One of the driving influences in the development of a popular fabricated salty snack was the great increase in the density of packaging due to the ability to stack the identical saddle-shaped pieces. The snack was intended to compete with potato chips made from potatoes, which have a relatively short shelf life and low bulk density. As a result of the low bulk density and short shelf life, potato chips are made close to population centers to reduce distribution costs. Fabricated chips, which are made from reconstituted dehydrated potato granules, are packaged in a sophisticated multi-layer canister to extend their shelf life by protecting them from oxidation. Their high bulk density and extended shelf life meant that the entire country's supply could be manufactured in one central location, resulting in economy of scale, compared with the dispersed, multi-site system of "regular" potato chips. Nonetheless, fabricated chips have only achieved about 5% of the total snack market. One suggestion for that performance is that they do not taste very good compared with the competitive product.

15.2 Size

The size of a new facility and hence its cost is probably the single most critical decision to be made. If a firm has several existing plants and is expanding essentially the same business into a new area, then it probably has a good idea of the capacity and capability of a plant of a given size. Many companies build almost standard facilities to accommodate growth and expansion. The more difficult challenge is when a product is new and there is uncertainty about needed capacity. In this case, it might be wise to build a minimum sized facility with provision for expansion in the future. This brings up some other issues that must be resolved.

15.2.1 Capability

As discussed earlier, a new plant is considered to accomplish a given mission in a given area. Executing that mission then involves a number of other decisions, including the following:

- Number of lines
- Warehouse or not
- Distribution center or not
- Expansion in the future
- Flexibility

The number of lines and flexibility are closely related but separate issues. Often a plant makes several products or families of products, perhaps various package sizes, or different manufacturing processes, as in candy where a plant might have starch molding, tableting, and panning. Sometimes processes share common facilities or equipment, such as a kitchen or mix room. Other processes may have common packaging lines, as in bagging, canning, or pouching. One challenge is to decide how independent any line or group of connected equipment should be. Will it be self-sufficient and able to operate without reliance on another line or will it share some equipment or services? A case can be made for either approach.

How to handle relatively low volume products, as distinct from high sales volume products, can affect design. It is usually desirable to minimize inventory, as discussed further later, but there also may be some minimum length run that makes sense for a given process. Essentially, a decision must be made about over what period of time products will be made. Each product may require some downtime and a changeover. Most operations try to minimize these and run for as long a period as possible, but shelf life, storage requirements, and customer demands can have opposing influences.

The more equipment is shared, the lower the investment and the higher the utilization of capital resources. On the other hand, sharing resources may mean that some lines cannot operate while others are in use, and this can limit the capacity of the plant. Ultimately, food process design usually results in careful compromises that are deemed to make sense for a given company in a given business. There is no one answer that fits all cases. A common solution is to have shared support facilities such as ingredient preparation and dedicated processing and packaging lines that are optimized for certain products and package sizes. In any event, the overall process design is chosen (Maroulis and Saravacos 2003, Clark 2005b, Lopez-Gomez and Barbosa-Canovas 2005).

It is good practice to standardize on a limited range of common process equipment, such as pumps, conveyors, instruments, and controls in order to simplify training, preventive maintenance, and to minimize the amount and variety of spare parts.

Other characteristics of the facility are then decided. Most food plants have separate warehouses for raw materials, finished goods, and packaging material. Raw materials and finished goods are separated to reduce the risk of contamination, since many raw materials are agricultural products that may be dirty. Depending on the overall plant layout, raw materials and finished goods may be handled in separate parts of the plant. Storage conditions can also vary, as in ice cream, where raw milk and cream are refrigerated, while ice cream is frozen. A major recall of peanut products in 2009 was aggravated in part by possible contamination of finished goods by raw materials.

Packaging materials should be treated as if they were ingredients because they come in contact with food and can be contaminated. Different materials have different requirements. Glass containers may be cold upon arrival in winter and need to be carefully warmed up to prevent condensation of water and thermal shock when they are filled with hot food. Paper for labels, cartons, and pouches needs to be stored in controlled humidity to acclimate to the plant environment and then perform properly in packaging equipment. Metal cans need to be protected from moisture to prevent corrosion. Plastic foil also needs to be stored in a temperature- and humidity-controlled environment, and all packaging materials need to be protected from dirt and other contaminants. In addition, because packaging materials are used toward the end of a typical line, it may be convenient to store them nearer to the point of use than at either end of a plant. Sometimes packaging materials have a dedicated receiving door. Soft drink plants often move cans directly from a truck to the line, in part because the cans are relatively bulky and are used in large quantities.

A modern plant tries to minimize the amount of area devoted to storage by carefully controlling inventory, using "just-in-time" purchasing and delivery practices and producing closer to order than to inventory. Finished goods storage can be minimized by promptly removing finished goods as soon as a truck can be filled. The products may go to customers, to a company DC, or to a customer or third party DC. One way to ensure that inventory is kept low is to not build much storage in the first place, but this may be resisted by plant operations people who would rather produce efficiently and not be concerned about whether they had made too much. Making to order may mean that less than a full shift of production is sufficient, which is a departure from traditional practice of scheduling full shifts, whether the product is needed or not. So, an early decision to make is how much, if any, finished goods storage to provide.

15.2.1.1 Examples of Inventory Decisions

A pet food company decided to build a new plant to make both wet (canned) and dry (extruded) pet foods. This was a major project, and, as almost always happens, the preliminary cost estimate soon exceeded the budget. There had been conceptual discussions about reducing inventory, but the preliminary design had a conventional finished goods warehouse. In an early cost-reduction meeting, the suggestion was made and accepted that no warehouse at all be built. Instead, there would be a small staging area big enough to accumulate a truck load or two and all product would be transferred to a DC as fast as a truck could be loaded. The savings in capital were significant. One suggested disadvantage was that potential expansion space was then not available, but, in fact, the proposed location of the warehouse was not ideal for expansion anyway. (Some designs do assume such future use.) The plant was built without a warehouse and operated successfully.

Another food company, manufacturing canned products, which typically have shelf lives of about 2 years, declared that it intended to reduce inventories drastically and to try and sell only products that were less than 9 months old. One consequence was that the company needed to find international sources for raw materials that had traditionally been seasonal products in the United States. This could add to costs because of increased transportation charges, but a much larger benefit was a great increase in equipment and facility utilization. Previously, most of a year's production was manufactured in a relatively short period following harvest of raw materials and then stored as finished goods. Much of the factory then was quiet and unused for many months. Under the new philosophy, the company was able to close one of its facilities, increase production, and avoid investment in a planned new facility, saving hundreds of millions of dollars.

For a multi-product company, a DC both receives and ships the full line of products. DCs are usually located to serve regional markets and may or may not be adjacent to a plant that produces some or all of the products. If a new site is also to be a DC, then it does not need a finished goods warehouse, but it will need additional truck docks to receive outside shipments and will now have a more complex order picking function. Especially when entering a completely new market, as in another country, it may be advantageous to construct a DC with a factory so that the market can be served initially from one location, with one management team. There is, of course, additional cost in building a larger facility, but warehouse space is generally less expensive to build than is factory space.

Capability of future expansion is almost always required, but how much must be decided. It is not uncommon to double the size of a plant over time. There needs to be sufficient land for such expansion and the design should anticipate where and how it will be done. Certain features of a plant are difficult and expensive to move, so these should be located on a face of the plant not designated for expansion. Such features include docks, labs, and offices. Provision for their expansion should also be made. Likewise, support facilities need not be fully sized for the ultimate capacity, but provision should be made for their expansion when necessary, typically by leaving extra space in various places, such as machine rooms, maintenance shops, and electrical switch rooms.

Flexibility is a profound topic. Here, in discussing initial design scope decisions, the impact is to consider whether the plant needs to make many short runs of various products in different size packages or longer, more standardized runs. Many choices flow from this decision. For example, a line with much of its equipment on casters that can be rolled into place or out as needed is very flexible, but may be less efficient than a large capacity, fixed line. A flexible line needs extra space to store equipment that is not in use. Flexible lines may be down more often for changeover and cleaning and so are less productive. However, a flexible strategy can reduce inventory and may be more profitable than a more dedicated line. Again, there is no single answer for all circumstances, but there does seem to be a trend toward more flexibility.

15.3 Overall Layout

Overall layout, the way equipment is located with respect to other pieces, is dictated by efficient flow of materials and people. The usual options, for one level, are

- straight through,
- U shaped, and
- L shaped.

Straight through layout has generally straight and parallel production lines in which raw materials enter one end and finished products emerge at the other end.

One advantage is relatively easy expansion by adding similar lines so long as there is sufficient space. Another advantage is that there are aisles between and alongside of the lines for people and fork trucks.

To minimize distances that materials or finished goods are transported, it is common that raw materials be received and stored at one end and finished goods be stored and shipped from the other end. This puts trucks on two sides of the building and requires two sets of receiving and shipping clerks and two areas for truck drivers to wait. As previously mentioned, it is not necessarily a disadvantage to have storage of raw materials and finished products separated, but there can be extra expense in building sufficient truck docks with their equipment (lights, levelers, locks, doors) in two locations. When all the doors are in one area, they can serve double duty for shipping and receiving. Having them all in one area also means an additional face of the building is free for potential expansion.

Arranging lines in parallel can be difficult if they are of unequal lengths. If they all start together, then some will end before others. This can result in wasted space and extra transport distances. Some processes can have parallel lines for part of the process, as in baking ovens, but must fan out when they come to packaging, which occupies a much greater floor area than do the ovens. In some bakeries, packaging is put on two or more floors so as to minimize the footprint of the building.

An alternative process layout is "U" shaped, where lines reverse their path so that they begin and end on the same side of the building. This allows for shipping, receiving, and storage to be in one general area, allows truck docks to be used more efficiently, and may save clerical and material handling labor. On the other hand, it is physically difficult to arrange similar lines this way and moving people along a line or between lines is more difficult.

A variation is "L" shaped, in which lines have one turn of 90°. This can result in a more square-shaped, rather than long and narrow, building. It has similar advantages and disadvantages to the straight through option, but can accommodate lines of different lengths.

Another issue is that of levels in a building. The least expensive way to build is to have one level or floor, but there are circumstances in which multiple levels can be advantageous. Certain types of food processes are traditionally multi-level, including flour milling, ready-to-eat breakfast cereal, and pet foods. These have in common that they use solid raw materials and produce solid products. Solids are easy to handle by gravity flow, so often these processes convey raw materials to a top floor by elevator or pneumatic conveying, then transfer between subsequent operations by dropping through chutes. Even meatpacking used to operate in multi-level buildings, relics of which still exist in Chicago's Union Stockyards. Animals walked up ramps to the kill floor at the top and primal cuts and by-products were conveyed by gravity until they emerged on the ground floor.

A multi-level building has a compact footprint, but bears extra costs for its structure, for the floors and for transporting of people and material among floors (elevators). They also can make communication among the workforce difficult in modern, automated plants where there may be relatively few people on each floor. Elevators, in particular, are seen as expensive and often unreliable, requiring maintenance and becoming a threat to interrupt production by breaking down. The common solution is to install multiple elevators, adding to cost.

A popular compromise in food plant building design is a high bay structure with equipment placed on a relatively open structural frame that is separate from the structure of the building. The frame has walkways and work places where people need to be, but is sufficiently open that the platforms do not qualify as floors, which can affect building code requirements for fire protection and also can influence tax assessments. There is often debate between making work platforms of open grid or solid plate. Solid plate is generally preferred because it prevents dirt from shoes or spills falling and contaminating food on lower levels. Where the food is confined and protected, open grid can be used for its lighter weight. However, it is hard to clean.

Partial mezzanines are another approach to achieving some of the benefits of multi-level and gravity flow without the disadvantages of a full floor. With a mezzanine, some equipment or storage bins are elevated, access is by stairs, and equipment can be located below. A common example is dry mixing, where the mixers are located above loading or packing stations. In considering any partial floor or mezzanine, it is important to understand local building and fire codes because these can affect exit door requirements, fire protection, insurance rates, and tax assessments. Companies want work platforms to be considered part of the equipment and not part of the building whenever possible because buildings are often assessed in part on their floor area. Equipment is also subject to more rapid depreciation than is a building for tax purposes. The space under a work platform needs its own lighting and fire protection. As a general rule, enclosed work spaces need two separate exits for people in case one is blocked in event of an emergency.

In summary, the process, the scale of equipment, and the flows of people and materials will dictate the three-dimensional arrangement or layout. This in turn will strongly affect the overall shape of the building. Other influences on the shape include the other functions that are present in addition to the process, the specific site, local requirements, and other architectural considerations. Lately, a desire to be "green" and sustainable may also influence building design.

Green design and LEED (low energy and environmental design) are efforts to reduce the energy and carbon footprints of a facility by being efficient in lighting, heating and air conditioning (HVAC), using recycled materials, and using other design elements as appropriate. Some features that contribute to sustainable design include the following:

- Lighting in warehouses that is controlled by motion sensors, so it is only turned on when people are present;
- Low solvent epoxy coatings;
- Heat recovery from oven and boiler stacks;
- Water reuse and recycle;
- Low inventory ammonia refrigeration systems; and
- The use of sustainable materials in office finishes.

It is difficult for a food plant to satisfy many of the goals because relative to commercial or institutional buildings, industrial buildings in general and food plants in particular are relatively energy intensive and have constraints on their design that limit the LEED features. Primary among these limitations is sanitary design.

15.4 Sanitary Design

Sanitary design refers to those building design features that may be unique to a food plant and are intended to reduce the risk of contamination by biological, physical, and chemical hazards (Imholte 1984, Clark 1993b, 2007, Jowitt 1980, Troller 1983). Some of the same principles are found in pharmaceutical plants. Hazard Analysis Critical Control Points (HACCP) refers to a rigorous approach to identifying, anticipating, making provisions to correct, and documenting potential hazards to people from food. It also provides a framework for food plant design. Food plants convert agricultural raw materials and other ingredients into edible and safe foods and beverages while protecting foods from spoilage and from posing a hazard to humans. Hazards may be microbiological, such as *Clostridium botulinum, Escherichia coli* O157:H7, *Salmonella*, or *Listeria monocytogenes,* organisms that can cause disease; or they may be physical, such as glass, wood, or metal; or they may be chemical, such as pesticide residues, cleaning chemicals, or unapproved additives. Critical control points are identified in the process, which if properly achieved, will control hazards. Procedures are devised and documented to respond to deviations from control points. These are primarily the concern of operations. The task of sanitary design is to minimize risks of contamination and to make easier the challenges of cleaning and maintaining the plant and equipment.

Since microorganisms require water and food to live, one approach to reducing their presence is to deny them these essentials. This leads to the principle that if a plant is normally dry, keep it dry because it is difficult to remove water once it is present. Plants that are normally wet anyway, because they process liquids, must be designed so that water does not accumulate and so that they can be easily cleaned. In practice, this means designing so that floors drain quickly and using materials and finishes that are resistant to cleaning chemicals, water, and heat.

Most food equipment is made of stainless steel because it is resistant to corrosion and can be polished so that food and dirt cannot easily cling to it. In addition, welds must be smooth, corners rounded, and the equipment designed so that it can be taken apart and inspected. These requirements follow from the second essential of microbial life – food. In the course of food processing, many foods form films on surfaces with which they are in contact. These films can harbor microbes. If the surface is rough, the film is difficult to remove and it may be hard to detect whether the surface is clean by inspection. Some types of equipment can be cleaned-in-place (CIP) by passing strong hot cleaning solutions through at high velocities, but inspection is still required to ensure that the cleaning process is successful. Often food equipment is designed so it can be disassembled without requiring tools or at the most one or two simple tools, such as a wrench.

Some equipment can only be cleaned by hand washing – clean out of place (COP) – in special sinks. The conversion from COP to CIP was one of the great advances in food plant design and permitted the growth in size of dairies, in particular, by reducing the labor needed for sanitation (Seiberling 1997).

For the same reasons stainless steel is used in equipment, it is often chosen for building mezzanines and work platforms in food plants. At first, this seems extravagant, but in the long run, it is a good decision. Stainless steel does not require painting, as would carbon steel structural members. Unprotected mild steel or carbon steel will corrode in the wet atmosphere of most food plants and the rust could become a food contaminant. Paint can chip and become a contaminant. Painted surfaces require recurring maintenance. Galvanized (zinc coated) carbon steel is an acceptable structural material, but poses its own challenges. Structural members can be galvanized in a shop, but when they are assembled by welding, the coating is removed at the welds, creating a location for corrosion unless the coating is replaced in the field. It is generally best to use stainless steel for structures in a food processing area. Some plants have gone so far as to use stainless steel panels for floors, walls, and ceilings. This may be a bit extreme for most cases, but is justified when conditions in the space are so severe and the food so vulnerable that the durability of stainless steel is required.

Doors, drains, pipe hangers, and even electrical conduit should be made of stainless steel in a wet environment. A common convenient structural unit, a perforated metal channel with the ends bent in, should not be used in a food plant because it is hard to clean. Contractors like it because it is familiar and easy to use to make pipe and conduit hangers. Instead of it, supports should be solid bars or angle members, with the angles pointed down so that the piece does not trap dirt. Pipe and conduit that runs along walls should be set about one inch away from the wall using special clamps or standoffs so that the space behind can be reached for cleaning.

In a dry plant, where dust is a concern, metal does not need to be stainless steel, but flat surfaces should be minimized. It is common to encase vertical columns in concrete bases with sloping tops, to make curbs at walls with sloping tops, and to cove or curve the intersection of floors and walls so that that area is easier to clean. Dust is a concern in a dry plant because it can harbor insects, attract rodents and birds, and be a potential explosion hazard. Many food dusts from flour, sugar, and starch are explosive in certain concentrations, which can occur in confined spaces, such as ductwork and equipment. A slight spark or static electricity can set off an explosion and the initial shock can create a second dust cloud, which may do even more damage.

Dust in a food plant is commonly controlled with a central dust collection system, which is a vacuum pneumatic system with connections to hoods over bag dump stations, mixers, and other locations where dust can be generated. There are one or more receivers with replaceable filters on the exhaust, which should discharge outside the plant. The filters must regularly be inspected and replaced. The discharge must be inspected and must comply with air emission regulations. Dust collected in the receiver is discharged to a bag or drum for disposal. The amount collected should be observed and the air flow in the system carefully balanced to achieve the desired compromise between elimination of dust in the air and yield loss. The dust collection system can exceed the explosive concentration limit in its ducts and receivers, and so all electrical equipment must be spark-proof and the entire system correctly grounded, so static electricity does not build up and create a spark. The dust collection system must be inspected and cleaned periodically because dust can build up in ducts to the point that the system is no longer effective.

Insects, rodents (rats and mice), and birds are concerns because they can spread microorganisms, insect parts are considered foreign matter in food, and their droppings are also contaminants. Evidence of rodent infestation is enough to cause a food plant to be shut by public health authorities. Designing for protection against rodents is a challenge because they are attracted by the abundant food present in food plant, they can go through very small holes, and they can gnaw through wood and some other materials. Food plants should have a rodent control program using traps placed by a professional and regularly serviced. There should be a clear path around the inside and outside perimeter of a food plant to allow placement of traps and to prevent rodents from being diverted into the plant by obstructions. Rodents are nearly blind and find their way by feeling surfaces with their whiskers, so if they encounter an obstacle along a wall they turn into the plant, where they can cause contamination. It is a continuing challenge to keep the perimeter paths clear, as it is tempting to use that space for casual storage. Normally the inside perimeter path is painted distinctively to remind people that it is not to be blocked.

Insects are controlled in part by devices that attract them with ultraviolet light and then kill them with electricity. Air curtains at doors to the outside are used to keep potential contaminants, including insects, from the outside from entering the plant when the doors are opened. An air curtain is created by a strong fan forcing air through a downward facing slot nozzle at the top of the doorframe. The fan turns on automatically when the door is opened. An air curtain can also help keep high humidity air from entering a freezer and thus reduce frost formation in the freezer. (Freezers should not have doors opening directly to the outside, but rather should have vestibules or small rooms between their entrances and the outside. These vestibules can be used as coolers.) If a space in a food plant becomes infested with insects, it may need to be fumigated by a licensed professional, using approved chemicals. Typically, even approved chemicals are toxic to people and great care must be taken to ensure that no residues enter the food. An alternative to the use of chemical pesticides is the use of heat to kill insects and their eggs. This is achieved by heating the space – which might be the entire plant – to 140°F and holding for 24 h. Such prolonged heating might damage computers and electronic controls unless they are protected with cooled enclosures or are shut down. Achieving such temperatures is usually beyond the capability of normal heating systems, so if heat sterilization is contemplated, the system must be designed accordingly.

Wood, paper, plastics, and glass are potential sources of foreign matter contamination and so should be minimized in food processing areas where food is exposed. Thus it is good practice to transfer materials from common wood pallets to captive plastic pallets for use within the plant. Better still is to have a separate space where bags and drums of ingredients are emptied into totes or other reusable and sanitary containers, keeping paper and plastic bags away from the food. This also separates a source of dirt – the outside of containers holding ingredients.

Glass should not be allowed in a food plant, which can mean minimizing windows or putting windows high on a wall (clerestories) so they are less likely to be inadvertently broken. When foods are packed in glass containers, care must be taken in material handling to prevent breakage. This means synchronizing conveyors and fillers so containers do not jam or contact each other. Samples are collected in plastic containers rather than in glass and mercury in glass thermometers are replaced with electronic instruments.

Lighting in a food plant may be fluorescent, high intensity halogen, or incandescent bulbs. Mercury vapor lights should not be used because they attract insects. Glass bulbs must be shielded so that if they break, glass is not scattered. Different lighting levels are needed in each area, ranging from 15 to 20 fc where the objective is safety, as in warehouses, to 40–50 fc where inspection occurs. Lighting spectrum should simulate sunlight so food appears most natural. For LEED or sustainable design, lighting can be controlled by movement sensors, in places such as warehouses, where people are not always present, so it is only on and consuming energy when people are present.

Floors, walls, and ceilings should be smooth and impervious so they can be easily cleaned and will not retain dirt and dust. A good material for walls and ceilings is fiberglass reinforced polymer panels. Insulated metal panels with enamel coatings are commonly used for partitions and sometimes for outside walls, but they are vulnerable to puncture by fork trucks. Concrete masonry units (CMU) coated with high quality epoxy paint are commonly used for structural partitions. It is important to use higher grade CMU with smaller pores and then to seal those before applying the epoxy. CMU should be laid so that seams between them are continuous in the vertical direction – an uncommon masonry pattern – but this permits water to drain easily and not get trapped in the intersections between units laid in the normal, overlapping pattern.

Floors may be made of brick, tile, terrazzo, or coated concrete. Brick was traditional for food plants and is still used in dairies and other operations where truck traffic is heavy and floors are often wet. The grout used to lay the brick must resist acid and basic cleaning solutions as well as extremes of temperature. Tile is less common and is suited to lighter traffic areas. Terrazzo, a monolithic concrete with aggregate filler, is durable and resistant, but relatively expensive. A common choice offering a good balance between cost and durability is a filled epoxy coating over concrete. The filling of aggregate in the coating (small stones or sand) is to adjust the thermal expansion of the coating to match that of concrete so the coating resists exposure to live steam, hot water or cold refrigerant. Otherwise, the coating can easily be stripped from the floor when exposed to extreme temperatures because of differences in the rate of thermal expansion between concrete and the coating. Paint or sealant is the least expensive floor treatment, other than leaving the concrete bare, and are rarely adequate for a food plant except in warehouse areas.

Wet areas need floors that slope to drains and enough drains to remove water quickly. Drains may be hub or trench and should be stainless steel until they join the main sewer. Hub drains are familiar circular openings covered with a grating or sometimes a plate with a perimeter slot. One drain per 200 ft^2 of floor area is common. Trench drains are shallow cuts in the floor covered with a removable grate, sloping to a connection to under floor drain pipe. Sanitary drains (from rest rooms) should be separate from process drains until they are outside the plant, where they may be combined in a grease trap that also serves as an air break, so that waste cannot flow back into the plant. Floating fats and oils and sinking solids are periodically removed and disposed of in landfills.

Trench drains can have high capacity for flow, but because of the extensive open grating, they need extra provisions for cleaning, because they can harbor microorganisms and spread these in aerosols. Aerosols – suspensions of small drops of water in the air – are created when hoses are used to clean floors and equipment. It is poor practice to use hoses while food is exposed because of the risk of contamination. One approach to cleaning trench drains is to connect the CIP system to the drain so it is cleaned as if it were another piece of equipment.

Ceilings are a frequently debated feature of food plants. Sometimes they are required to completely enclose a space to control temperature, humidity, odor, or dust. In buildings where the roof is supported by trusses, a ceiling over processing areas protects against dirt and dust from the hard to clean metal supports and from leaks. An ideal roof for a food plant uses precast double tees, which need not be painted and have few surfaces on which to catch dirt and dust. A new plant can be specified to have precast construction, if it is available in the area, but many existing buildings have less expensive metal trusses, or more rarely, wooden trusses. Trusses are nearly impossible to clean and so a ceiling of fiberglass-reinforced panels may be needed. The cost of pre-cast concrete construction varies with area of the country due to differences in supply, competition, and demand.

The disadvantage of a ceiling is that the space above it is often forgotten and can become dirty. The remedy is routinely to inspect and clean the top of the ceiling by removing some panels for access. Lighting and fire protection is usually required above and below the ceiling, depending on local codes. Some ceilings are made of panels that are strong enough to support people walking on them. This permits easy access for servicing piping and ductwork running above the ceiling.

Spaces that may pose hazards such as dust explosions or excessive noise (compressor rooms, boiler rooms, hammer mills) often have special construction requirements, including blast proof doors, blow out panels to the outside, and reinforced CMU walls. These requirements may be dictated by local codes.

Since raw materials are a major source of potential contamination, it is important to separate raw materials and their containers from exposed finished product. There is also increasing concern about allergens and preventing allergenic materials from contaminating foods in which they do not belong. Separation is achieved by partitions, by balancing and controlling pressure and airflow, and by controlling access by people. The most common food allergens of concern are as follows:

- Peanuts
- Wheat

- Milk
- Eggs
- Tree nuts
- Fish
- Shellfish
- Soy

Ninety percent of food allergies are associated with these categories. People are protected by labels that declare whether a food contains known allergens or whether it was made in a plant that also processes allergens. It is a serious matter to produce a food containing allergens without properly declaring their presence on a label. The number of people avoiding certain ingredients is increasing, primarily due to greater awareness of food allergies.

Airflow should be counter to the flow of product so that it contacts clean product first. People may be issued distinctively colored uniforms so that it is easy to see if they are where they belong. Hand washing and footbaths are often used before entering an area where food is exposed. Often overlooked is the fact that maintenance mechanics enter and leave all parts of a plant and so may transmit contaminants. They must be trained to keep themselves, their tools, and their vehicles clean. It is common to use color-coding for brushes, shovels, and other tools so those used for raw materials or inedible waste are not used to contact finished products or edible rework.

Finally, one of the most important principles of good sanitary plant design is to allow enough space for safe movement and easy inspection. It is always difficult to justify space in the design phase because equipment dimensions are rarely well known and so initial layouts seem poorly filled, but as time goes by and details are filled in, space always gets tight. At the same time, costs almost always rise and the easy way to cut costs seems to be to reduce size. This is usually a mistake.

15.5 Security

In any manufacturing facility, there is concern for employee safety, prevention of theft, and prevention of sabotage. Food manufacturing raises the additional concerns of food safety, prevention of deliberate contamination as well as inadvertent contamination, and the safety of visitors. Thus there are good design and operational practices that address these concerns. Sadly, this is an evolving area, so existing plants may need to revisit their conditions and practices. New plants need to anticipate possible changes in the future.

Training and education of employees is the first priority. This is a never-ending task as there can be high turnover in food manufacturing. Employees should know who belongs where, should wear their picture identification and challenge anyone who does not, and should personally observe safety and access rules. Employees should understand that a risk to their company is a personal risk to them, of not having a job, at least. Past instances of deliberate contamination have often been attributed to disgruntled employees. Thus, good employee relations can be considered a first line of defense. This need not mean abandonment of discipline nor of sound business practices in compensation, but it does mean that investment in employees can be preventive of serious problems.

Security in design includes such features as a perimeter fence, gates at vehicle entrances, card-controlled access to interior spaces, limited access for truck drivers, and control over visitors. One consequence of limiting access for truck drivers is the need for a rest room and lounge where they can wait while a truck is loaded or unloaded. A perimeter fence prevents casual trespassing, which carries the risk of liability for injury to a visitor – even one without permission to visit. Fences can be easily penetrated and so are not the sole precaution taken against unauthorized entry. Access to the plant should be limited and guarded.

Visitors need to show identification and are usually escorted by an employee. Contractors and consultants need identification, are instructed in plant policies, and may be limited in where they can go unescorted. Visitors and employees remove jewelry (except for plain wedding bands), remove anything from pockets above the waist (to prevent objects from falling into food), and wear hair and beard nets. Visitors and employees usually wear company-issued smocks, which may be color-coded. Smocks do not have buttons, which can fall off, but rather use snaps or ties. Shoes should be closed, and in some plants may have steel safety toes. Some plants require bump hats or hard hats because they may have low equipment or other potential hazards.

Food manufacturing, especially meat packing, can be relatively hazardous because some tasks are repetitive, sharp tools are often used, and there are many pieces of moving equipment, such as conveyors. Workers' tasks should be analyzed for repetitive motion risks and people cross-trained so they can be rotated away from making the same motion for too long. Visitors should be cautioned against touching or contacting equipment or food, except where specifically permitted. Tobacco, other food, and glass containers are not usually permitted in a food plant. A break room/cafeteria is usually provided, commonly equipped with vending machines, but sometimes with catered food. Smoking is only permitted outside of the plant. In other countries, several free, hot meals may be provided each day to all employees. While seen perhaps as an amenity, there is also a safety and security aspect to providing a place for eating and relaxation. In any event, provision for such a space must be provided.

In developing a food process and specific plant layout, it is important to identify each place where food is exposed and vulnerable to deliberate or accidental contamination. Measures that can taken include securing hatches and entrances to tanks, covering conveyors, directing video cameras at such locations (and then monitoring the images!), and having responsible employees at such locations. People are the best observers of quality in foods, so inspection by visual observation as well as by instrumentation and laboratory analysis is needed at many points in a process. Such inspectors and observers are also guardians as well.

As with the HACCP plan, where the needed response to a deviation in the process is determined ahead of time, it is important to anticipate and prepare for breaches

of security or safety. For example, employees need to know what to do and where
to go in the event of a fire or severe weather. Where tornadoes occur, a place in
the plant needs to be designed for and designated as a shelter. There needs to be a
rallying point in the event of an evacuation, as may be required if ammonia leaks
from a refrigeration system. There needs to be coordination with local authorities
for response to fire or other emergency. Some plants employ nurses for care of minor
injuries, in which case a room needs to be provided for supplies and treatment. The
value of training was illustrated by the 2009 safe landing of an airliner in the Hudson
River, in which the pilot and cabin attendants used their skill to preserve the lives of
155 passengers and crew.

Fire officials need to know if there are any hazardous chemicals on the premises,
such as tanks of oil or areas where dust accumulates. Supervisors need to be trained
to account for people in an evacuation and to safely shut down equipment in an
emergency. Often plants have a public address system for paging, which can also be
used for announcements and instructions, but food plants can be too noisy for these
to be heard well. Many plants also have a radio system for contacting individuals.
Closed circuit video monitoring is useful to observe operations, reduce pilferage,
and prevent tampering. Obviously, it is only useful if the screens are watched.

15.6 Support Facilities

Support facilities refer to the areas of a plant used for purposes other than direct
manufacturing. These include space and equipment for utilities, space for employees
to change and eat, and offices and labs.

Common utilities include the following:

- Steam
- Electric power
- Compressed air
- Water
- Fuel
- Sewer
- Refrigeration
- Hydraulic fluid
- CIP chemical
- Nitrogen or carbon dioxide

Steam is usually raised in a boiler. Good practice is to have at least two boilers,
each capable of providing about 75% of the required total load, so that the plant can
operate, perhaps at reduced capacity, if one boiler is down for maintenance or suffers
a failure. Boilers for food plants normally produce steam at 100–150 psi, which is
reduced as needed at the point of use. It is rare to need much higher-pressure steam
in food processing. Steam is normally distributed in carbon steel pipe with threaded
connections. A good practice installs steam traps at low points in the lines to remove

condensate. Discharge of steam traps should be direct to drains or to condensate return systems. Condensate is recovered both to save heat and to save specially treated water. Boiler feed water may be softened to reduce scale formation, which costs money. Steam and condensate in carbon steel pipe picks up rust and corrosion products, so if discharged to the floor, it will quickly stain the floor.

Culinary steam is specially generated for direct contact with food. It is common to inject chemicals into boiler feed water to prevent corrosion and scale formation in the boiler and piping, but these chemicals are not intended for food use. Culinary steam is raised in a heat exchanger using high-pressure steam to heat purified water. The culinary steam is filtered and piped in stainless steel to where it is used. To minimize the cost of stainless steel piping, culinary steam is normally generated near where it is used.

Boilers are often fired with natural gas, but may use coal, fuel oil, or waste material. Fuels other than natural gas may require special boilers. Boilers that use wastes, such as paper, wood or crop residues, need special controls to compensate for the variable heating values of such fuels. With the rising costs of fossil fuels and the increasing emphasis on sustainability, it may become more common for food plants to use unconventional fuels in steam boilers. A high pressure, stationary boiler usually requires a licensed operator in attendance. This can be a specially trained member of the maintenance staff who may have other duties. Steam below 15 psi does not require a licensed operator and is often adequate for many food processes.

Other heat sources for processing include closed loop hot water and hot oil systems. These use steam or fuel to heat a circulating medium, such as water under pressure (so it can exceed 212°F) and heat-stable oils, which can exceed 500°F. Circulating fluids may be preferred over steam because they do not pose the potential hazards of steam and, in the case of oil, can reach higher temperatures than are practical with steam. They also can provide good temperature control in processing equipment. While boilers are usually located in a relatively central place, circulating systems are usually located close to the use point to minimize distribution costs and are considered part of the process.

Boilers should be located on an outside wall with access to outside air for combustion and with room to pull tubes if necessary. Often, a large louvered panel is used both for air intake and potential access. An air emissions permit is usually required for the boiler stack. Most natural gas does not require treatment of its combustion products before discharge, but coal or waste fuels may, depending on their sulfur content and the efficiency of the boiler.

There are situations in which a food plant can get steam from an outside source, such as a co-generation plant or a neighbor with excess steam. This can save on capital, but makes the plant dependent for an important resource on another party. It might be wise in such a case to install a stand-by boiler or to establish in advance a source from which to obtain a replacement boiler. There are firms that promise to deliver a boiler on a trailer in less than 24 h.

Electricity is usually supplied to the site by a high voltage line from the local utility. The plant needs to install a step down transformer and then an electrical distribution system with several voltage levels – 440, 220, and 120 V are common.

The higher the voltage, the less expensive the wiring and the more efficient the conversion of electricity to useful work in motors on equipment or to light. Equipment and areas of the plant are usually wired separately so individual pieces can be isolated. Switchgear and motor starters generate heat and so should be in rooms that are ventilated. Much of the switchgear is only rarely accessed and so can be located relatively remotely.

Lockout procedures must be established and people trained in their use for safe maintenance and operation of the equipment. Only those authorized should touch electrical controls. Each person is issued a lock and key and is responsible for personally shutting off power to a device on which he or she intends to work. All moving equipment should have emergency stop controls (E stop) in easy to reach locations.

Control and communications wiring is normally relatively low voltage and should be isolated from power wiring. Wire in a food plant is normally contained in metal conduit, which protects it from accidental damage, water, dust, insects, and rodents. Conduit should not be overfilled. Local codes usually dictate good practice. Conduits should not be installed too close together for ease of cleaning. Conduit should be one inch away from walls, for the same reason, and hangers should be of the same sanitary design as is used for utility and process piping, that is, no perforated channel, no all-thread rod. In some plants, putting wire in cable trays is acceptable, but these are difficult to clean. The benefit is lower cost and easy access.

With the importance of computers to modern manufacturing plants, it is important that a power source be reliable. One approach is to have power supplied from two independent sub-stations so that at least one will always be available. This is not always possible to arrange. Another approach is to have a stand-by electrical generator fueled by natural gas or diesel fuel. Often, the generator is supplemented by a large battery uninterruptible power supply (UPS) that detects a drop in incoming voltage, supplies power temporarily, and switches to an alternate source.

Compressed air is almost always required in a food plant because it is used in many controls and instruments. Air compressors are noisy and so should be isolated, often in the same room as boilers and refrigeration equipment. Compressors also generate heat and saturate compressed air with moisture, so compressed air should be cooled and dried. Air to be used in instruments should be oil-free, meaning an oil-free compressor is used or there is a filter on the compressed air to remove lubricating oil mist. Intake for air should be directly from the outside to avoid compressing conditioned air (if it were taken from inside the plant) and to reduce noise. Compressed air systems normally have a storage tank to buffer surges in use. Typical pressures are about 100 psi with reduction at use points. Compressed air is usually distributed in carbon steel pipe with threaded fittings, but may use copper or plastic tubing after reduction.

Water is important in food processing as an ingredient, for cooling and for cleaning. Water must be potable, meaning suitable and approved for drinking, but even potable water may require additional treatment to remove chlorine and suspended matter. As previously mentioned, boiler feed water is often treated by ion exchange to remove hardness (calcium and magnesium salts that can form scale). Water for cooling may be reused either for cooling again, by returning to a cooling tower, or for other purposes, such as cleaning. Water in a closed cooling circuit may have chemicals added to reduce scale and corrosion, in which case care must be taken that such water not contaminate food.

Water may be supplied by a public or private utility or from a well or surface water source, such as a lake or river. If the firm supplies its own water, then it is responsible for testing and treatment to ensure that it is potable and suitable for processing. Water quality often varies with seasons and weather, though water from an underground source varies less than that from other sources.

Most of the water that enters a plant eventually is discharged to a sewer and then to the environment. The exception is water that is used in the product. If this is substantial, it is prudent to meter incoming and discharge water separately and to negotiate fees based on actual use. Otherwise, it is common to base sewer charges on incoming water rates on the assumption that these are approximately equal. Water may be distributed in carbon steel pipe, plastic pipe, or copper tubing. In process areas, ingredient water should be distributed in stainless steel tubing.

Fuels were discussed earlier in regard to the boiler. Fuels may also be used in ovens, dryers, and some other equipment. The normal choice is natural gas because it burns cleanly. Where natural gas, which is mostly methane, is not available, lique- fied petroleum gas (LPG), which is mostly propane or butane, may be used. Natural gas is supplied by underground pipe from a utility company. LPG is delivered by truck to above ground pressurized storage tanks. Fuels are normally distributed in carbon steel pipe to use points, at which the pressure is reduced as needed. There is a difference in heating value between natural gas and LPG, so burners need to be designed for the specific fuel. For direct contact with food, as in ovens or dryers, it is uncommon to use other fuels, but wood and coal have been used, and for special purposes, such as smoking meats, wood (as saw dust) is preferred.

Sewer service was discussed previously in considering sites. The preferred approach is to discharge to a municipally owned and operated treatment plant. Both sanitary (from rest rooms) and process waste water can normally be handled by such plants because food process wastes are biodegradable, though they may be strong, meaning higher in BOD and suspended solids than conventional sanitary wastes.

Inside the plant, sanitary and process wastes should be kept separated in their own cast iron pipes. Underfloor drains are combined outside the plant in a grease trap/air break. This is a chamber in which the separate drains enter at a high level and the connection to the treatment plant is made at a lower level. The chamber permits fats, oils, and grease (FOG) to float to the top, from which they are periodically skimmed by specially equipped trucks and disposed of as solid waste. Heavy solids may sink to the bottom, from which they are removed as needed. The separation of inlet and outlet in the chamber construction prevents wastewater from flowing back into the plant and possibly contaminating food and equipment.

In some plants there is a need for a central hydraulic power system. This is a network of carbon steel pipe in which a special oil is circulated under high pressure and then returned after use to a supply tank, from which it is pressurized and used again. The high-pressure oil is used to drive special motors and other equipment by reducing its pressure through small turbines. It is important that the system be kept very clean, as tolerances in the motors are tight. Hydraulic power has been used where there is a lot of water, creating the risk of shorts if electric motors were used, and for driving conveyors where speeds are continuously modulated, simply by varying the flow of hydraulic fluid to each motor.

Where one or more central CIP systems are used, there may be central storage and distribution of cleaning chemicals, which is another support system. Common cleaning chemicals are sodium hydroxide also called caustic soda, hydrochloric or nitric acid, and a sanitizer, which might be an iodoform, a quaternary ammonia compound, or another substance. Caustic and acid are received as concentrated liquids, which are hazardous and corrosive. They are stored in plastic or stainless steel vessels surrounded by concrete dikes, in case of leaks or spills. Chemical storage should be separate from other areas and access restricted to those properly trained and with a need for access. Concentrated chemicals are delivered to use points in plastic or stainless steel pipe. They may diluted in line or after arrival. The use points are small tanks with pumps and associated piping for a section of the process. Properly diluted acid and caustic are circulated through the piping and equipment, normally returning to the tanks for reuse after filtering and replenishment of used chemicals.

Nitrogen or carbon dioxide may be used for cryogenic freezing or for blanketing storage vessels, as inert gas. Either gas is normally delivered as a pressurized liquid by truck and stored in highly insulated steel or stainless steel vessels. Often the gas vendor supplies the storage vessels. The liquids are very cold, so distribution lines must be insulated to reduce condensation and ice formation. If only used for blanketing, typically of oil in storage to reduce oxidation, it might be cost efficient to install a nitrogen generator, which recovers nitrogen from atmospheric air by using special membranes. In cryogenic freezing, the liquefied gases allow rapid freezing at very low temperatures.

15.7 Welfare Facilities

Welfare facilities refer to those spaces devoted to supporting the people and equipment in a food plant. Among these are the following:

- Lockers and rest rooms
- Machine rooms
- Maintenance shop and parts storage
- Equipment storage
- Break and eating space
- Meeting, conference, and training rooms

In addition, most plants have other spaces for offices and laboratories, including the following:

- Executive and management offices
- Team rooms

- Sales, accounting, and personnel offices
- Shipping and receiving offices
- Quality laboratories
- Research and development laboratories
- Nurse's station
- Company store

The area devoted to these functions is determined by the number of people served or occupying, by corporate standards, and by functional requirements. For example, each employee should have a full length locker if they are required to change clothes before going to work, as many food plant employees are. The locker rooms need also to have showers, space for clean and used uniforms, hand washing stations, and toilet stalls.

It is common to provide one or two cots in the women's locker room.

Machine rooms house refrigeration compressors, boilers, air compressors, pneumatic conveying blowers, electrical switchgear, and perhaps telephone and computer equipment. Often such equipment is noisy and can give off heat. These spaces are ventilated, but usually not air conditioned, except for more delicate equipment such as telephone and computers. It is helpful to have overhead doors to the outside for ease of installing and removing the often large and heavy equipment.

A boiler room needs a source of combustion air, which should be from the outside, often through louver panels, which can be removed for access. Likewise, it is best to supply air compressors and pneumatic blowers with outside air, rather than plant air, which is often conditioned – heated or cooled. Air compressors should have after-coolers and oil filters because a common use is for instruments and air-activated controls. Pneumatic blowers also should have after-coolers unless the heat of compression is considered useful. Space over machine rooms is often used for storage because height requirements are usually less than elsewhere in the plant.

A maintenance shop should be equipped with appropriate machine tools, welding equipment, small parts storage, and tool storage. Control of parts and tools can be challenging, so access to the shop is usually restricted to those authorized. It is also a dangerous place for those not properly trained. Larger equipment storage may be off-site unless the equipment is frequently used. Modern food lines often have easily replaced units that are rolled in/rolled out as needed. These need to be put somewhere safe and convenient.

There was a time when it was common for food plants to provide hot meals on every shift and in many international sites it is still common. In the United States, the trend is towards vending machines serving snacks, soft drinks, hot drinks, and some prepared foods. An outside vendor, who then needs daily access, commonly services the machines. Smoking is no longer permitted in most food plants, so employees who wish to smoke must go outside to a designated area. If this is provided, it should be sheltered so it can be used in inclement weather.

Spaces for management and other support personnel can range from strictly functional for a medium scale plant to quite opulent for a corporate headquarters adjacent to a plant. It is still common in US firms for the space allowed for an office, and the quality of furnishings, to be commensurate with rank – larger and fancier for those with more responsibility. However, there is a counter-trend in which offices are more egalitarian or even eliminated altogether in favor of open arrangements or bullpens, some without even having partitions. The choice is a matter of corporate culture and usually has been established well before a new or expanded plant is considered. However, the exercise of considering the design of such space could be the impetus for re-evaluating the usual assumptions.

With or without executive and management private offices, there is always a need for conference rooms and spaces in which visitors, vendors, and employees can meet privately. These are usually equipped with speakerphones, white boards, computer network connections, tables, and chairs. A good practice is to have several such rooms of varying size, ranging from one in which most employees can be accommodated to one suitable for a meeting among two or three people. Increasingly, companies are organizing their work force in teams with a relatively flat structure, as distinct from the traditional hierarchy with several levels of supervision. Essential to the team approach is providing a space in which each team can meet, work together, and communicate. These spaces are not available as conference rooms because the team needs to be able to leave work in progress on white boards, store papers, and references and otherwise leave and return as needed.

Mid-level, as distinct from management or executive, personnel work in sales, accounting, human resources, and some other functions that need work space, normally cubicles defined by partitions. They have phones, computers, and some storage. A common mistake is to make these spaces too small, with inadequate allowance for inevitable expansion. There are rules of thumb about how much space is sufficient for these functions. Suppliers of modular office furniture often provide the service of designing such spaces in the hope of selling furniture.

Shipping and receiving needs one or two small offices, depending on whether shipping and receiving are together or separated. There is also a need for a small lounge where visiting truck drivers can wait while their vehicles are loaded or unloaded. The lounge should have a small toilet room so the drivers need not go into the main plant.

Quality assurance needs a small laboratory where typical tests can be performed. The exact space and furnishings depend on the specifics of the plant – volume and variety of products, tests that are required, and number of technicians. Many quality tests are performed on the line by operators using simple equipment such as scales, but other tests, such as fat content, moisture, and salt need more conventional laboratory space and equipment. In addition to lab benches, cabinets for storage, and instruments, the lab space needs work space for the technicians to hold a phone, computer, and files. Further, there needs to be space to hold retain samples for at least the life of the products plus some safety allowance. Typically, samples are retained from every lot.

Separate from the quality lab should be any research and development (R&D) space that is required. Often all R&D is performed at corporate headquarters or another facility, but some plants, especially those that are the sole plant of a firm, may have R&D on site. It is important that R&D not interfere with the normal operation of the plant, so it is helpful to have a pilot plant as well as a development lab. A pilot plant is a small space with small-scale equipment that is flexible and can simulate the full-scale production processes. Pilot plants can be expensive investments but they pay for themselves by permitting development work to proceed without using production scale equipment, over-time labor, and excessive amounts of raw material.

Finally, many plants have a company store for employees and sometimes for the public. Practices vary by company, ranging from fairly elaborate stores that stock a company's full line, to small counters where employees can pick up designated quantities of locally produced products. Sometimes the store is used to dispose of off-spec but safe product, perhaps mislabeled or close to the end of its code date. In some industries, it is a long-standing tradition that workers are entitled to a given amount of product each month.

This brief description of the spaces and functions in a new plant is not meant to be all-inclusive, but rather to inspire thinking about these issues early in a project, where they can be accommodated efficiently and allowances made in the budget. Too often they are not considered and then are either ignored or designed poorly. Properly considered and designed, they contribute positively to morale, safety, and quality.

15.8   Discussion Questions or Assignments

1. Pick a food company from the top 100 (see the August issue of *Food Processing*, Putman Media, Itasca, IL). Where do they have facilities now? Where in the United States might they go next? Where internationally? Why?
2. Pick a state or region. What incentives do they offer plants that might locate there?
3. What does raw land in your area cost per square foot? What does developed industrial land suitable for a food plant cost? What do existing industrial buildings of about 100,000 ft^2 cost?
4. Find the costs of utilities in your area or another that might be assigned or of interest. Do costs change with volume of use? Are there shortages or surpluses?
5. Where does water come from in your area or another of interest? What is the water analysis? Would it need treatment for use in a typical food plant? In a beverage plant?
6. What fuels are available in your area or another area of interest?
7. Pick a company of interest. Describe the corporate culture. How might it influence the design of offices in a new food plant?
8. Pick a food product. What are the required quality tests? How are they performed? What equipment is needed? Can any tests be done on line? Can they be done automatically and in line?

Chapter 16
How to Tour a Food Plant

Whether as a guest from the outside, as a consultant, or as an employee from headquarters or another site, touring a food plant is an opportunity to be taken seriously and seized as a teaching moment. Common sense and good manners dictate that you introduce yourself to your host and to others you may meet. Each plant may have its own safety and sanitation requirements, but most will require that a visitor wear a distinctive badge, take off jewelry (except simple wedding bands, usually), wear a clean smock, a hair net (and beard net if necessary), wear closed toe shoes, and ear plugs or other hearing protection. In some plants, hard hats may be required, and some issue rubber boots.

Usually there is a briefing in which the purpose of the visit is discussed and an overall description of operations may be given. It is common to tour in the same direction that materials flow, from receiving raw materials to shipping finished goods. Some portions of a plant may be considered so proprietary that an ordinary visitor may be excluded from those areas. In other plants, where allergens may be processed, visitors may be excluded from certain spaces to prevent cross-contamination.

So, subject to the usual restrictions, what should you look for?

16.1 Material Handling

How and where are ingredients delivered and stored? What are they? Are they inspected on receipt? Are they segregated until approved? What happens to rejects? Same questions for finished goods, when you get there. How and where are packaging materials received and stored? How are ingredients and packaging materials delivered to lines?

If the hoppers holding powders or particulate materials are dented from being hit to induce flow, you can safely conclude there are or have been serious material handling problems. These are often opportunities to make a constructive contribution, if that is your role.

Note the major methods of moving material around the plant. Some common ones include

- Pneumatic conveying for solids (pressure or vacuum)
- Fork trucks for pallets, drums, and bags
- Intermediate bulk containers (totes, big bags, Gaylords – disposable totes)
- Conveyor belts for containers, cases, and some foods
- Conveyor chains for cans and other containers
- Bucket elevators
- Screw conveyors
- Freight and personnel elevators
- Pipes, often stainless steel tubing, for liquids, gases, and utilities
- Drains in the floor for wastewater – should be stainless steel in process areas. May be trench or hubs (round or square). Do the floors slope to drain?
- Ducts for supply and removal of heated or cooled air
- Hoods for dust collection

Note the provisions for electric power, control wiring, lighting, and security. Are wires contained in metal conduit, in cable trays, or are they loose? Is the conduit mounted away from the walls to allow cleaning behind it? Is the conduit neatly installed?

16.2 Other Sanitary Design Features

Normally a visitor is not expected to be critical or conduct an audit, unless that is the purpose of the visit, which it can be. However, the knowledgeable visitor notes conditions and forms an opinion, which may or may not be later shared.

The overall morale and management style may be portrayed in the cleanliness and neatness of a plant. A food plant should be clean, with minimal water or product on the floor. Depending on your status and responsibility, you might or might not comment on your impressions and offer suggestions. Use good sense and diplomacy. Running a busy plant with too few people – the most common case – is hard.

Look under equipment. Can it be reached for cleaning? Are the tops of equipment cabinets dusty or greasy? Does equipment have extraneous panels covering areas that must be accessed?

Are painted areas peeling? How are the floor and walls finished? Is the equipment running smoothly or are there frequent interruptions? There usually is an optimum speed for any line, but there is constant pressure on plant people to operate faster. Eventually, the optimum rate is exceeded, as indicated by frequent stops. The correct response is to reduce operating rate until the time between stops is significantly longer. Net production almost always increases at the lower optimum rate because there is less down time and better quality. It makes no sense to produce a poor quality product faster.

16.3 Characterizing the Plant

Can you sketch the process flow diagram after your tour? How many products and processes are made and used? Is the plant wet or dry? Refrigerated? Dusty? Does it feel crowded? Has it been expanded? Can it be further expanded? What are compatible processes or products?

How are wastes handled? What are the wastes? Are efforts made to reduce waste? Almost any plant can reduce its water use by such methods as recycling cooling water, controlling flow from hoses, and reusing cleaning water.

How is air flow controlled in the plant? Is there a risk of cross contamination from raw to finished goods? Are there differences in room pressures? (You can tell by how hard it is to open some doors.) Are there air curtains at doors from the outside?

Did you notice a rodent control program? There should be professional traps around the inside wall. Is there a clear path around the inside wall? Often there is one when the facility is built, but it gets filled over time with tools, boxes, and equipment. Are there traps and a clear path around the outside wall?

Is there an insect control program, such as UV traps? What type of lighting is used? Are outside lights mounted on the building or at a distance? Mercury vapor lamps attract insects and are best used away from the building. Rodent and insect control programs should be provided by professional experts who are properly licensed.

How are the processes controlled? It is increasingly common for food processes to be controlled through computers and graphical human machine interfaces (HMI). So far as you can tell, do the processes reflect the latest developments in the appropriate fields? It is expensive and not always necessary to continuously upgrade equipment, but it is a sign of visionary management when selective investments are made in process improvements.

Where are the people in the plant? Do they seem busy? Are they performing physical labor or are they mostly watching and tending operating equipment? Are many inspecting product? It is increasingly common for line operators to be responsible for quality assurance on the line. There often is a special bench where pictures of gold standard products are posted with computer terminals for entering data.

Are there visible distinctions among workers, such as color coded smocks? This is often done to indicate where workers are supposed to be, to identify maintenance workers, and to identify visitors. Do workers have picture ID cards? Do doors require codes or electronic key cards to open? Are there sanitizing hand wash stations and footbaths at entrances?

16.4   Some Examples for Discussion

1. A dry baking mix plant in Europe was built as a multi-level tower to take advantage of gravity flow, with staging on one level, mixing on the next, surge bins on the next, and packaging on the lowest. The equipment was modern and automated, but the mixer chosen was difficult to empty completely, which poses a challenge when changing formulas. More significant, however, was the fact that each floor was made of open metal grate. The concept, evidently, was to permit air flow through the floors so there could be a common dust collection system. Grate is also lighter than a solid floor of poured or pre-cast concrete. However,

the open grate permitted dirt from workers' shoes to fall through to the level below, posing a real threat of contamination.

2. In one very large plant, the cafeteria seemed always to be full. Upon inquiring, it developed that there was an aggressive union and generally hostile relations between workers and management. Efforts at discipline had been unsuccessful. A new plant manager succeeded in discharging a small number of persistent malingerers and suddenly productivity improved and the cafeteria seemed less full except at lunch.

3. A plant making ingredients from agricultural materials experienced high microbial plate counts in its air and products because the raw materials were dumped close to the packing line. It is moving packaging to a separate room and erecting a wall between the dump area and the rest of the plant.

4. The common practice in several large canned soup plants is to receive and wash potatoes, onions, carrots, and mushrooms in a separate room, well removed from the mixing and filling areas.

5. A large canned pet food plant had no refrigerated storage for its meat and poultry ingredients. Instead, unprotected blocks of frozen meat were stacked on the bare concrete floor. Blood, fat, and pieces of meat were all over the floor, making walking challenging. In contrast, many other pet food plants are as clean and safe as most food plants. Pet food plants should be built and operated like human food plants because pets are susceptible to many of the same hazards as humans are and, inadvertently or deliberately, some pet foods are consumed by humans.

6. In all the plants of a major snack food company, workers on each line stop every hour and carefully evaluate and taste the products they are making, comparing them to pictures of target product for color and common defects.

7. In two modern sea food plants in Central America, the rooms are refrigerated, raw material receiving and preparation is separated from further processing, and workers wear caps, smocks, gloves, and rubber boots. They wade through deep sanitizing basins in and out of each room and wash their hands at each entrance.

8. Many bakeries traditionally had wooden floors, in part to deaden the noise of steel wheeled dough troughs being moved. If the floor got wet, they often warped and required repair. Further, flour and sugar easily got beneath the boards and provided harborage for insects. Most of such floors are being replaced with epoxy coated concrete.

9. A candy plant installed in an existing warehouse had an overhead pipe dripping a liquid into a bucket next to a starch molding machine. The same plant had loose wiring, peeling paint, cobwebs, and standing water on the floor. Ironically, the plant operator was in a dispute with the building owner over allegedly leaking skylights (among other issues).

10. In a vegetable cannery in Eastern Europe, six women were chopping cabbage at a metal table with kitchen knives, no head covering, and wearing aprons.

16.5 Lessons

1. Much can be learned by careful observation during a tour of a food plant.
2. Material handling is at the heart of many food processes and greatly influences the design.
3. The state of worker morale and management effectiveness can often be discerned by observation of housekeeping and worker behavior.
4. A visitor should remember he or she is a guest and should offer comments, if any, tactfully and appropriately.

Chapter 17
Build New, Expand, or Upgrade?

It often falls to an engineer in a food company to evaluate options for adding capacity to manufacturing. He or she will rarely handle this task alone, but instead will be part of a team that may also include some outside specialists, such as architects, engineers, and consultants. This chapter will summarize some of the steps that are usually taken, referring to several sources that give more detail (Clark 2005b,a, 1993a,b,c). The intent is to present here some checklists and procedures for making proper evaluations and considering significant factors.

Commercialization is the complex series of tasks involved in moving a product from concept to profitable manufacture (Clark and Levine 2009). In spite of its critical importance, there are relatively few references on the broader topic of product development and almost none on commercialization. There are many skills required, which means that commercialization is almost always a team effort. Further, many of these skills are not routinely a part of the typical food science educational experience, which means that commercialization is an interdisciplinary effort. Activities such as the common, senior-level student product development projects give a sense of the challenge, but the teams formed for such projects rarely involve students from engineering, marketing, or the other disciplines that are routinely involved in real life. Learning how to lead a commercialization effort and how to make a valuable contribution to the team is probably best achieved by participation in multiple teams with increasing levels of responsibility.

17.1 Do We Need a New Facility?

The options for manufacturing are to build a new facility, expand an existing facility, or use a comanufacturer. Before assuming that a new facility is needed, it is important to be assured that existing facilities that might be appropriate are fully utilized. Many food plants operate 5 days a week with two shifts, using the third shift for cleaning and sanitation. Other types of manufacturing and, increasingly, many food companies, approach 24/7 operation, that is, round the clock every day – or nearly so. Thirteen days of operation, with the 14th used for maintenance, is not uncommon.

J.P. Clark, *Case Studies in Food Engineering*, Food Engineering Series,
DOI 10.1007/978-1-4419-0420-1_17, © Springer Science+Business Media, LLC 2009

Many food products are seasonal because of the limited availability of raw materials, meaning that the equipment may be idle much of the time. Companies are increasingly addressing this issue by obtaining raw materials from other places, such as the opposite hemisphere; using partially processed raw materials, such as bulk aseptically stored fruits and vegetables; and using controlled atmosphere storage to extend the season for root crops, apples, and some other materials.

A new facility may be justified for several reasons: new technology may not fit in existing facilities; existing facilities may be at capacity (or may not exist); and/or logistics may dictate a new location close to markets or raw materials. New technology often requires a new facility simply because it occupies more space than is otherwise available. A good example is the automation of almost any task on a food processing line. New technology may also dictate a different layout than the machine it replaces. For instance, a bakery that converts from batch ovens to a conveyor oven would need a radically different arrangement. A facility to irradiate foods needs a highly specialized design with shielding and conveyors that are unique compared to any other facility in the food industry. A hydrostatic retort for continuous processing of canned food requires a very different layout than a comparable battery of batch retorts.

High capacity and highly automated food processes may not be appropriate for every situation. Developing countries need good jobs as much as they need anything, and labor is usually relatively inexpensive. Sophisticated equipment needs skilled maintenance and support, which is not evenly distributed around the world. Older technology is often robust, and used equipment is frequently available, lowering the capital cost.

Facility location is critical and may be dictated by sources of raw materials, by distance to new or existing markets or by distance to new or existing distribution centers. The relative bulk density of raw materials and products and the shelf life of products often determine whether a product should be manufactured close to markets or can be made farther away. For example, frozen and dehydrated potato products are commonly made near potato-growing areas, whereas potato chip snacks are made closer to markets. Products that have frequent, store-door deliveries are made close to market. Examples are soft drinks, refrigerated dairy products, most salty snacks, and bread.

17.2 Project Phases

The five normal phases of a facility project are as follows:

- Feasibility study
- Preliminary design
- Detailed design
- Construction
- Commissioning.

A feasibility study is intended to confirm the economic attractiveness of a proposed project. The feasibility study requires sufficient scope description and design so that costs can be estimated within about 30%, which is reflected in the contingency allowance. Normally, the proposed selling price, anticipated volumes, and unit variable costs are known from the product definition. The key remaining question is what investment is required to manufacture the product in the desired volume and location.

Conducting a credible feasibility study usually requires some outside assistance in the form of architects, engineers, consultants, and people with construction experience. The company needs to assign a team led by an executive with authority to commit the firm. Typically, members of the team include staff members from research, engineering, and operations. Staff from marketing, finance, human resources, information technology, and logistics may be involved in the study at times; they must be kept informed and their viewpoints considered.

Preliminary design should be performed only after the feasibility study confirms the desirability of investing in a new, expanded, or converted facility. This phase adds detail to the design started in the previous phase. Much of the work is performed by outside resources, such as an architect/engineering firm (A/E). A preliminary design takes longer and costs more than does the associated feasibility study. A specific site should be identified because site-related factors are some of the larger causes of uncertainty in cost estimates. The scope description prepared in the feasibility study is expanded in this phase, but the basic assumptions should not be changed.

Detailed design is the phase in which construction documents – drawings and specifications – are prepared. Naturally, it involves more people, typically engineers for each discipline, and costs more than preliminary design. During detailed design, one of the critical functions is the resolution of potential interferences. Interferences occur when two disciplines try to use the same physical space in the plant for some purpose. One of the major responsibilities of the project manager is to identify and resolve potential interferences quickly. It is much less expensive to resolve these on paper than to do so during construction. A useful strategy is to pre-assign chases or designated paths and areas in the plant for various purposes. A good practice is to have most utility piping outside a process area with only vertical drops to use points.

Construction is normally the responsibility of a general contractor (GC) or construction manager (CM), but the owner is ultimately responsible for safety, costs, and quality. A GC may perform some tasks himself, but primarily hires subcontractors in the various trades and disciplines. He makes a profit by marking up the costs of labor and materials from the subcontractors.

The more complete the scope definition and the design documents, the fewer the changes. Sometimes owners are the source of changes, but they should be aware of the consequences, which almost always are to increase the cost of a project. If an owner wants more transparency and control, then the owner can engage a CM, who normally charges a lower fee or profit margin than a GC does, because the CM assumes much less cost risk.

Commissioning begins when construction is complete and consists of operating the process with realistic raw materials, but with no expectation of high yields. The objective is to find and correct deficiencies in the process and equipment, fine tune conveyors, and train operators. The objective of identifying a commissioning phase is to budget time and money for these essential tasks.

17.3 Equipment Design, Selection, and Scale-Up

Food companies rarely design processing equipment. Rather, they or their consultants usually select equipment from existing lines offered by numerous vendors. The key documents in selecting equipment are process flow sheets, material and energy balances, and a process description.

When evaluating processing equipment vendors, it is important to confirm that they observe industry standards. Cost, quality of components and fabrication, service capability, and delivery time are other factors that affect the choice of one vendor over another.

The final stages in commercialization are demonstration in a pilot plant and plant trials. Plant trials are expensive and should only be performed when there is a high probability of success. It should be obvious that commercialization is a complex exercise involving people from a wide range of disciplines and requiring excellent communication and cooperation.

17.4 Examples

1. A major breakfast cereal company observed that one of its competitors had built a relatively new plant in the South Central part of the United States. They wanted a new plant of their own in the same general area, but beyond the reach of the local union. Accordingly, they chose a site in an adjoining state. Because of changes in the market, the plant capacity was not needed for a few years after completion and so it stood vacant. During design, there had been debate over whether to use a proprietary piece of equipment that was very effective for one product but could only be used for that purpose. The alternative was to install older but more versatile equipment. The specialized equipment was selected, but years later a company engineer told the designer who had argued for flexibility that it would have been a better decision to choose the versatile equipment. It turned out that the site chosen was in a flood plain. Fortunately, that was learned before any equipment was installed and possibly damaged by water.

2. A pet food company had traditionally designed and built its plants based on feed mill technology – a common practice in the industry. While evaluating engineering firms for a planned new plant, they heard the argument that pet food plants should be designed and built to human food plant standards. This concept was so appealing, even though it meant higher investment costs, that the engineering firm proposing the approach was chosen for the new plant assignment.

Subsequent work from the same client over several years made the relationship one of the more successful the engineering firm ever had.

3. A major ice cream company needed capacity to service the Southern California market, but wanted a location outside of the expensive Los Angeles basin. A site in the lower Central Valley was chosen. An important constraint on the site was a limit on the amount of wastewater that the local municipality could handle. Water conservation became an important consideration. Through a variety of techniques, including putting valves on hoses, limiting the number of hoses, selecting fillers that used less water than other possibilities, and recovering and reusing rinse water, water use per gallon of ice cream was cut by a factor of ten compared with their other plants.

4. A fresh produce distributor in the Midwest felt it had outgrown its urban, leased facility. A produce distribution center is basically a group of refrigerated rooms maintained at various temperature and humidity levels. Some products need to be separated from others because they give off ethylene, which affects the ripening of other fruits and vegetables. There were differences of opinion among the owners over desirable locations – urban or suburban – influenced by traffic, labor supply, and available incentives. Over 50 existing buildings and sites were evaluated. Existing buildings tended to be larger than needed, but with fewer truck docks than desired. Available land controlled by real estate developers was usually capable of supporting a larger facility than was immediately needed. In the end, the owners decided they really did not have enough money for a new facility and, besides, they thought they could grow the business without additional space by shipping directly to customers from their suppliers rather than bringing shipments to a central facility.

5. A candy plant was moved from the United States to Canada. Some existing equipment was relocated from the previous site, some was moved from other sites, and a small amount was purchased new. The plant did not work very well. Without quite realizing it, the company had introduced a completely new process for the most critical operation. Further, they had moved the older equipment "as is." For reasons that are not entirely clear, older equipment that seemed to be working just fine, when relocated and reinstalled, always seems to fail. When moving an existing operation, the equipment should be carefully evaluated before moving to determine whether it should be used. If it is worth moving, it should be first sent to a shop where it can be rebuilt and restored. It is much easier to do this type of work in a shop than in a working factory. Before modifying a process, as by introducing a piece of equipment that was designed for another purpose, the new process should be tested and optimized, preferably in a development environment.

17.5 Lessons

1. Commercialization is a multi-discipline team activity. Few individuals have the experience and skills to do it alone.
2. How to provide additional capacity requires a complex analysis of markets, site features, and other factors.

3. There are few, if any, ideal potential sites because most of the good plots of land have already been used. This means that almost every potential facility site will have some challenge that will require creativity and money to conquer.
4. It can be faster to convert an existing building to a food plant than to build a new one from a scratch, but it is not necessarily less expensive to convert because older buildings almost always need some renovation.
5. Likewise, moving older equipment almost always requires significant investment in maintenance and repair before reinstalling the equipment. Such work should be done in a shop with appropriate facilities, not on the factory floor.
6. Processes can be modified in profound ways by introducing a new piece of equipment for a new function. The new process should be tested and optimized in a development environment, not in the plant.

Chapter 18
Developing Processes

Processes are developed to manufacture new products or to improve manufacture of existing products. It is common for many new products to use existing plants and process lines if possible in order to minimize capital investment. Companies often speak of platforms meaning technologies that underlie several products. Examples are baking for cookies and crackers, frying for snacks, extrusion for breakfast cereal, and aseptic processing for food service dairy mixes. Even within a platform, there can be development challenges for new applications.

Developing truly new processes is rare but can be satisfying and rewarding for the firm that succeeds. One approach is described below.

Morphological analysis is a technique related to Synectics and some other ways of organizing ideas and stimulating creativity that has been used successfully to develop novel processes. The concept is to identify the key characteristics or parameters of a candidate challenge and then to identify the various options for satisfying each parameter. Existing solutions, if any, are represented by some path through the matrix of options and parameters. Paths not taken may be novel solutions. It is analogous to creating a meal from a Chinese menu that has multiple columns of choices. It works.

New food processes may be as simple as replacing one unit operation with another or one brand of equipment with another. On the other hand, a new process may apply a fundamentally new technology, though those are pretty rare. In any case, it is necessary to understand what the rate limitations of a given sequence of operations are. A good practice is to set a line speed by the most critical operation, but it is not always clear what that is (Clark 2005b).

Food processes must always be concerned about food safety, which means applying the principles of sanitary design to both equipment and plant. These are discussed in detail in other sources, but will be summarized here.

18.1 Sequence of Process Development

It is common for food scientists to focus on product development and engineers on process development, but in reality it is difficult to separate the two activities. Ideally, both disciplines as well as other people in a firm, such as those from marketing, plant operations, and distribution should be involved as a team. Phases in a development project are often defined by the scale of market test that is conducted. Market test results serve as guides to the people who are gates in a stage/gate approach. The objective is to minimize risk and to identify potential failures as early as possible. Historically, food product development is difficult with a surprisingly high failure rate, even for products that reach national markets.

A typical sequence is as follows:

- Concept development (often by marketing using focus groups)
- Prototype manufacture (bench top/kitchen, often by food scientists and chefs)
- Internal testing (taste panels, informal cuttings, executive reviews)
- Cost estimate/formula revision (may be first involvement of engineers and operations)
- Manufacture of small quantities for central location tests (CLT). A CLT typically occurs at a shopping mall or other location where random samples of consumers can be recruited and their reactions recorded for analysis. (Manufacture may be done in a pilot plant or at a third party facility. It involves engineers, food scientists, procurement, and relationship managers for third parties.)
- Formula revision if necessary based on test results or abandon project if results indicate likely failure of concept.
- Manufacture of quantities for home use test (HUT). A HUT involves providing test samples to consumers for use in their home, with reactions obtained by survey or interview. Larger quantities and more realistic packaging are necessary, which means package development has been proceeding in parallel. (The same facilities previously used for manufacture may be used or a new one sought. The same company people are involved. There may be a little more attention paid to cost at this stage.)
- Formula revision as necessary. Package and instruction revision may be appropriate. Abandon if results suggest likely failure and no easy remedy. (Involves decisions at management level.)
- Manufacture for multiple city test markets. Larger test markets are aimed at measuring effectiveness of advertising, pricing, positioning in stores as well as trial and repeat purchase on a reasonable scale. The test has to be long enough to permit repurchase to occur, which may depend on the type of product. Test market cities are typically medium sized and cleanly served by television stations because television advertising is a major vehicle for foods. There needs to be good overlap between chain store service areas and television network distribution. Manufacture needs to be near commercial scale, which means that the manufacturing strategy must be determined and prepared. (People involved include engineering, marketing, distribution, and operations. Manufacture may be in company facilities or by third parties.)
- Review of test market results and decision to proceed or abandon. Involves strategic level executives. Revision of product design is unlikely at this point. Manufacturing strategy may be revised – quantities raised or lowered, for instance.

- Product introduction. Typically, some time is needed to fill the distribution channels. Product introductions often are in the Fall as a new television season starts, schools resume, and people return from vacations. Project schedules are fairly rigid because advertising time is purchased far in advance, packaging material has a long lead time, and any new equipment required also can have a long lead time. Responsibility for manufacturing and distribution are typically assumed by operations, logistics, and procurement, with continuing support as needed from research and engineering.

Points to observe about this sequence include the numerous occasions when the project may be abandoned. Costs escalate significantly with each step. Good development management seeks to identify potential failures as early as possible. It sounds brutal, but there simply is not enough money nor resources to carry every idea to fruition.

The process developer needs to start early in the project visualizing how the new product or family of products will be made. Does it need a new platform or can it be made on one of the company's existing platforms? Does it need new equipment, even if it is using an existing platform, because all the available capacity is being used? Does it need new space? Does it make sense to use a co-manufacturer? Especially if a new platform is involved, it may make sense to use a third party because they will have relevant experience and less company capital will be at risk.

If a technology that is new to the company is involved, what is the best way to learn what is needed? One approach is to hire people with relevant experience and skill. Another is to use a consultant as a source of knowledge. Still another is to educate existing people as needed, by performing research in house, by attending short courses, and by consulting experts at universities and government laboratories.

A sure test of whether knowledge of a potential process is adequate is the ability accurately to model performance of the process (Maroulis and Saravacos 2003). While the various levels of market tests are occurring, the developer can be assessing resources, identifying potential equipment, and co-manufacturers and gathering knowledge as necessary.

18.2 Examples

1. A manufacturer of custom vitamin mixes was building a new facility to upgrade the operation. The existing process involved batch mixing, testing, adjustment of the mixture, and filling into cans. Some issues included the relatively low precision of some assays and the frequent loss of stored product because of age. Order quantities rarely matched batch size, so the excess product was stored, but was often of use only to one customer because each customer had its own formula. A morphological matrix was constructed in which columns were such features as batch size, method of formulating, method of mixing, and method of filling.

Rows are entries for various options of each feature (as an exercise, construct the matrix). One path through the matrix involves literally making one can at a time. The analogy was drawn to mixing paint in a hardware store. Relatively uneducated clerks can prepare one gallon of any color paint desired and exactly reproduce it as necessary. The new process was named blend in can (BIC) and drew upon technology from the largest manufacturer of paint blending equipment to fabricate an industrial implementation (Clark 1997b). Some advantages include reduction in inventory and consequent losses because everything is made to order, reduction in testing because formulas were prepared by precision dispensing, and greater automation. Subsequently, it was learned that some flavor mixtures are prepared in similar robotic equipment that was developed independently. The more general idea of assembling things one at a time may have wider application, in pre-plated meals, for instance.

2. In the early 1970s, a large baking company conceived the idea of fortifying snack cakes to provide a nutritious breakfast to school children. There were significant technical challenges involved in adding protein to a cake mix. Protein sources, such as soy concentrate, need to have off-flavors removed, and the cake crumb has to be made strong enough to support a nutritious, but not very functional, in a baking sense, ingredient. The result was a product that, with a glass of milk, was the nutritional equivalent of a full breakfast of bacon, eggs, orange juice, and milk. The educational and health benefits of giving undernourished children a good meal to start the day were undisputed. However, the company was severely criticized by "health activists," who claimed the project was merely a scheme to get children eating sweet goods in the morning. The company's defense was that the food would not be eaten if it did not taste good. The fortified food was not a commercial success at the time, but one could say that it was just ahead of its time and was a predecessor of energy bars, breakfast bars, and other fabricated foods delivering a high concentration of nutrients conveniently.

3. Many canned soups are made much as they are at home, but on a larger scale. Vegetable soup, for instance, may have 19 ingredients ranging widely in amount used and in physical form. These are typically prepared from raw agricultural materials that are washed; sorted; peeled (if necessary); cut into pieces; portioned (weighed) into barrels or buckets; added to a jacketed kettle; heated to disperse starches, soften texture, and remove air; and then filled into cans. The cans are cooked in retorts, cooled, labeled, and packed in cases. In developing concepts for a possible new plant, the suggestion was made to mix and fill cold, saving time and energy. Retorting might need to be extended because thermal process time depends on the initial temperature, but it was asserted that flavor would be improved because less would be removed during the kettle heating step. The process was adopted for some products.

4. A snack cake manufacturer wanted to substitute high fructose corn syrup (HFCS) for sucrose in some of its products in order to reduce costs. HFCS is almost always less expensive than sugar and many other large users of sugar had made the substitution, in soft drinks, for instance. One target product is a cream filled sponge cake that is almost an American icon. The first question that innocently

arose was, "Which formula shall we use?" "Wait, isn't there one national formula?" "No, turns out that each of nearly 20 bakeries has their own recipe." When informed of this situation, executives were skeptical, but after a side by side tasting of samples from the field, they realized that corporate headquarters had lost control of this product and, probably, of most others. A new recipe was developed using HFCS and installed as the national standard. An interesting side issue was the fact that the product from one bakery consistently had a grainy or sandy cream filling because crystalline sugar was not completely dissolved in their process. The bakery defended this defect as being what the market wanted, ignoring the fact that the market had no choice. HFCS produced a filling that was smooth, as intended.

5. Canned pet foods are made by mixing together various types of meats and cereals, grinding them, filling into cans, and retorting. The formulation is typically done by counting frozen blocks of meat that are meant to have a specific weight. Cereal ingredients are typically scaled and transported pneumatically. For a new pet food plant, and later for an existing one, an automated process was conceived in which meat ingredients are ground separately and then stored in agitated bins on scales. A computer containing batch recipes calls for the appropriate bin to discharge into a collecting screw conveyor. The mix is transported to a grinder/mixer and then to the filler. Other ingredients are added as needed in the batch mixer. The system is more accurate than the previous approach and involves less manual labor. Some challenges involve dumping large totes or bins of raw materials, keeping frozen meat from sticking back together, and controlling flow from the bins. In one arrangement, the bins were supported above the collecting screws to take advantage of gravity and to save space. The basic concept arose from asking why could not meat be portioned and moved as the cereal ingredients were. Other processes might benefit from similar thinking. (Can you think of examples?)

6. A baking company posed the challenge of developing a long shelf life yeast raised donut. Yeast raised donuts are essentially fried bread, but in many bakeries they are made from purchased mixes to which only water need be added. Yeast raised donuts stale very quickly, so donut shops claim they will not sell in the afternoon a donut made that morning. Commercial, nationally distributed snack cakes need a shelf life of at least 5 days to be practical. Yeast raised donuts fail by becoming firm in texture and by moisture from the body of the donut dissolving the common sugar glaze or icing. Without divulging the proprietary details, a solution was found by observing that the donut needed to be at its peak of quality when eaten not when first made. Many foods are tasted and processes optimized for qualities straight out of the oven or fryer, ignoring the fact that most foods change continuously over their lives. The quality of most foods tends to decline with time, but some actually improve with age. Think of some cheeses, wine, whiskey, brandy, and some confections (liquid center chocolate covered cherries, for instance). We wondered if a donut could be made that got better with time, for at least a while. We found a way that works. Along the way, we discovered that many of the ingredients in commercial mixes were canceling each other so

that both could be removed, saving cost. As an exercise, think of some other situations in which this insight about changes in quality over time might apply.

18.3 Lessons

1. Process and product alternatives can be systematically identified and analyzed using the tool of morphological analysis to concisely document existing solutions and discover new ones.
2. Solutions in food processing can often be found in other areas of technology. It pays to be observant and broadly knowledgeable.
3. Challenge each step of existing processes and each ingredient of existing formulas. Steps and components creep in over time and their justification may be forgotten or have gone away.
4. Think about where and when a product will be used and evaluate accordingly.

18.4 Closing Note

This book began as a collection of case studies for possible academic use, but it has morphed into a kind of technical memoir capturing some of the lessons I have learned over a long and varied career. I hope it is both useful and interesting to read. I am grateful to the many colleagues and clients from whom I have learned valuable lessons and facts.

Appendix
Glossary of Some Terms Used

Adiabatic. Thermodynamic term referring to the energy generated within a system as distinguished from energy transferred in or out of the system.

Angle of repose. The angle that a pile of powder or particles forms with a horizontal surface. It is a characteristic of the powder.

Dry bulb temperature. The temperature measured by a thermometer with its sensing element exposed to the environment.

Extrusion. The process of forcing a paste or dough through an opening to form a shape that is affected by the opening and may be cut off into pieces.

Finisher. A cylindrical screen, usually with a rotating paddle inside that wipes the screen and forces pulp or juice through, retaining seeds and skins of fruits or vegetables.

Half-products. Extruded pellets of cereal dough that are relatively dense and stable and can be stored and held easily. When needed, they can be expanded by frying or gun puffing to make porous snacks or cereals.

Homogenizer. A high-pressure pump that forces a liquid through an orifice to create small, stable oil droplets in an aqueous phase.

Humectants. Substances that attract water and help to control water activity and texture in foods.

Induction. The creation of an electric field by the action of an external alternating magnetic field.

Leavening. The creation of carbon dioxide within dough or batter by the action of yeast on sugars or by chemical reaction of an acid with sodium bicarbonate.

Load cell. A measuring device using a strain gauge to measure weight, usually of a vessel.

Mechanical vapor recompression. A type of evaporator in which steam from the fluid being concentrated is compressed to raise its temperature and used to supply the heat of evaporation. It uses electrical energy instead of thermal energy to drive off water from solution.

Micelle. Small structure of lipids and proteins in an emulsion that stabilizes a suspension of oil in water or water in oil.

Microbial ecology. The complex community of microbes – bacteria and other single-cell organisms – in a given environment.

Microwave. Electromagnetic radiation that interacts with water molecules in foods to generate heat.

Ohmic. Refers to heating by electrical resistance.

Overrun. In ice cream, the relative amount of air that is incorporated in a mix, expressed as percent.

Parenteral. Refers to nutritional solutions that are injected into a vein or artery instead of taken by mouth. Used in patients who are unable to use normal food.

Pulsed electric fields. Heating method in which electric fields are imposed in short bursts. Found to be an effective pasteurization technique for some foods.

Radio frequency. Electromagnetic radiation in the frequency range used for radio; imposed on foods as another technique for preservation.

Ribbon blender. Cylindrical vessel with horizontal shaft around which is wound a twisted blade to mix powders and doughs.

Screw conveyor. Device in which solids and pastes can be transported by rotation of shaft with flat band wound about it.

Screw feeder. Device which meters solid powders by rotation of shaft wound with flat band.

Sheeting. Process in which material is reduced in thickness by passing between rolls. Used to reduce thickness of dough for cookies, pasta, biscuits, and other cereal products.

Slide gate. Device to control flow of powders by changing the cross-sectional area of outlet from bin.

Standard of identity. Legally defined composition of certain foods. Aimed at preventing consumer fraud by defining what is meant by certain food descriptions.

Stator. The part of a pump that remains stationary or fixed while a shaft turns within it.

Surge. Provision made in a process for receiving material that cannot proceed further because of a process interruption or provision made to supply material to a portion of a process when the normal supply is interrupted.

Vibratory conveyor. Material handling device for transporting powders by moving a tray back and forth at a frequency and amplitude that does not damage the material but gives it a net forward motion.

Weigh belt. Measuring device for solids flow that includes a belt conveyor over a series of load cells that can measure the weight on the belt at any moment.

Wet bulb temperature. The temperature measured by a thermometer or other instrument whose sensor is surrounded by a water-wetted wick in equilibrium with the environment. The temperature is lowered by the evaporation of moisture, which is related to the relative humidity of the environment.

Bibliography

Altomare, R. E. 1994 Heat Transfer in Bakery Ovens, in *Developments in Food Engineering, Proceedings of the 6th International Congress on Engineering in Food,* T. Yano, R. Matsuno, and K. Nakamura, Eds., Blackie Academic & Professional, (an Imprint of Chapman & Hall), London.

American Meat Institute 2004 *Final FDTF Principles & Expanded Definitions*, 2 page summary, American Meat Institute, Washington, DC.

Barbosa-Canovas, G. V. 2005 Food Engineering, in *Encyclopedia of Life Support Systems*, UNESCO Publishing, Paris.

Bartholomai, A., Ed. 1987 *Food Factories Process, Equipment Costs*, VCH Verlagsgesellschaft mbH, Weinheim, Germany

Blanshard, J. M. V., P. J. Frazier, and T. Galliard, Eds. 1986 *Chemistry and Physics of Baking*, Royal society of Chemistry, London.

Burstein, D. and F. Stasiowski 1982 *Project Management for the Design Professional*, Whitney Library of Design, New York.

Charm, S. E. 1971 *Fundamentals of Food Engineering*, 2nd ed., AVI Publishing Company, Westport, CT.

Clark, J. P. and C. J. King 1968 An Improved Freeze-Drying Process Using Convective Heat Transfer, in *Food Technology 22*, Institute of Food Technologists, Chicago, IL, p. 1235.

Clark, J. P. 1987 Food Manufacturing: Opportunities for Improvement. *Food Technology 41* (12), 56–58.

Clark, J. P. and K. K. Fallon 1988 Where to Begin and How to Proceed: Automating Facilities Management Functions. *The Journal of the American Institute of Plant Engineers* September/October, 19–25.

Clark, J. P. and W. F. Balsman 1989 Computer Integrated Manufacturing in the Food Industry, in *Proceedings of 5th International Conference on Engineering and Food (ICEF 5)*, Elsevier, London, W. E. Spiess, Ed. and *Engineering and Food*, Vol. 1, Elsevier, London 1990, pp. 781–789.

Clark, J. P. 1991a Engineering and Manufacturing, in *Food Product Development*, E. Graf and I. S. Saguy, Eds., Van Nostrand Reinhold, New York, pp. 91–103.

Clark, J. P. 1993a Plant Design – Basic Principles, in *Encyclopaedia of Food Science, Food Technology and Nutrition*, Academic Press Ltd, London, pp. 3605–3608.

Clark, J. P. 1993b Plant Design – Designing for Hygienic Operation, in *Encyclopedia of Food Science, Food Technology and Nutrition*, Academic Press Ltd, London, pp. 3608–3613.

Clark, J. P. 1993c Plant Design – Process Control and Automation, in *Encyclopaedia of Food Science, Food Technology and Nutrition*, Academic Press Ltd, London, pp. 3613–3617.

Clark, J. P. 1997a Cost and Profitability Estimation Chapter 13, in *Handbook of Food Engineering Practice,* E. Rotstein, R. P. Singh, and K. Valentas, Eds., CRC Press, Boca Raton, FL, pp. 537–557.

Clark, J. P. 1997b Food Plant Design – Food Engineering in Practice, in *Engineering and Food at ICEF7,* R. Jowitt, Ed., Sheffield Academic Press, London, pp. AA42–AA48.

Clark, J. P. 1999 Food Plant Design and Construction, in *Wiley Encyclopedia of Food Science and Technology,* 2nd ed., John Wiley & Sons, New York, pp. 946–953.

Clark, J. P. 2005a Food Plant Design in *Food Engineering,* in *Encyclopedia of Life Support Systems,* G. V. Barbosa-Canovas, Ed., EOLSS Publishers/UNESCO, Paris, pp. 683–696.

Clark, J. P 2005b Food Process Design, in *Food Engineering, Encyclopedia of Life Support Systems,* G. V. Barbosa-Canovas, Ed., EOLSS Publishers/UNESCO, Paris, pp. 697–706.

Clark, J. P. 2007 Food Processing Plant Design, in *Encyclopedia of Agricultural, Food and Biological Engineering (EAFE),* Taylor & Francis, New York.

Clark, J. P. 2009 *Practical Design, Construction and Operation of Food Facilities,* Elsevier/Academic Press, New York.

Clark, J. P. and L. Levine 2009 Commercialization and Manufacturing, in *An Integrated Approach to New Food Product Development,* H. R. Moskowitz, I. S. Saguy, and T. Straus, Ed., Taylor & Francis Group LLC, New York.

Connor, J. M., R. T. Rogers, B. W. Marion, and W. F. Mueller 1985 *The Food Manufacturing Industries,* Lexington Books, D. C. Heath and Company, Lexington, MA.

Connor, J. M. 1988 *Food Processing an Industrial Powerhouse in Transition,* Lexington Books, Lexington, MA.

David, J. R. D., R. H. Graves, and V. R. Carlson 1996 *Aseptic Processing and Packaging of Food,* CRC Press, Boca Raton, FL.

Fast, R. B. and E. F. Caldwell, Eds. 2000 *Breakfast Cereals and How They Are Made,* American Association of Cereal Chemists, St. Paul, MN.

Geankoplis, C. J. 1993 *Transport Processes and Unit Operations,* 3rd ed., Prentice-Hall, Inc., Englewood Cliffs, NJ.

Goldblith, S. A., L. Rey, and W. W. Rothmayr, Eds. 1975 *Freeze Drying and Advanced Food Technology,* Academic Press, London.

Goldratt, E. M. and J. Cox 1986 *The Goal: A Process of Ongoing Improvement (rev.),* North River Press, Croton-on-Hudson, NY.

Goldratt, E. M. 1990 *The Haystack Syndrome,* North River Press, Croton-on-Hudson, NY.

Goldratt, E. M. 1994 *It's Not Luck,* North River Press, Great Barrington, MA.

Hendrickx, M. E. G. and D. Knorr, Eds. 2002 *Ultra High Pressure Treatments of Foods,* Kluwer Academic/Plenum Publishers, New York.

Holdsworth, D. and R. Simpson 2008 *Thermal Processing of Packaged Foods,* Springer, New York.

Ibarz, A. and G. V. Barbosa-Canovas 2003 *Unit Operations in Food Engineering,* CRC Press, Boca Raton, FL.

Iglesias, H. A. and J. Chirife 1982 *Handbook of Food Isotherms: Water Sorption Parameters for Food and Food Components,* Academic Press, London.

Imholte, T. J. 1984 *Engineering for Food Safety and Sanitation,* Technical Institute of Food Safety, Crystal, MN.

Jowitt, R. 1980 *Hygienic Design and Operation of Food Plants,* Ellis Horwood Ltd., Chichester, England.

Karel, M. and D. B. Lund 2003 *Physical Principles of Food Preservation,* 2nd ed., Marcel Dekker, Inc., New York.

King, C. J. and J. P. Clark 1969 *Systems for Freeze Drying.* U.S. Patent No. 3,453,741.

King, C. J. 1970 *Freeze-Drying of Foodstuffs,* CRC Critical Reviews in Food Technology, CRC Press, Boca Raton, FL.

Komolprasert, V. and K. M. Morehouse, Eds. 2004 *Irradiation of Food and Packaging,* American Chemical Society, Washington, DC.

Kozempel, M., N. Goldberg, and J. C. Craig, Jr. 2003 The vacuum/steam/vacuum process. *Food Technology 57* (12), 30–33.

Lewis, M. J. and T. W. Young 2002 *Brewing*, 2nd ed., Kluwer Academic/Plenum Publishers, New York.

Loncin, M. and R. L. Merson 1979 *Food Engineering: Principles and Selected Applications*, Academic Press, London.

Lopez, A. 1975 *A Complete Course in Canning*, The Canning Trade, Baltimore, MD.

Lopez-Gomez, A. and G. V. Barbosa-Canovas 2005 *Food Plant Design*, CRC Press, Boca Raton, FL.

Mannapperuma, J. D. 1997 *Design and Performance Evaluation of Membrane Systems*, Chapter 5 in Valentas et al. 1997.

Maroulis, Z. B. and G. D. Saravacos 2003 *Food Process Design*, Marcel Dekker, New York.

Maroulis, Z. B. and G. D. Saravacos 2008 *Food Plant Economics*, Taylor & Francis Group LLC, Boca Raton, FL.

Martin, C. C. 1976 *Project Management: How to Make It Work*, AMACOM, New York.

Matz, S. A. 1988 *Equipment for Bakers*, Pan-Tech International, McAllen, TX.

McCabe, W. L. and J. C. Smith 1976 *Unit Operations of Chemical Engineering*, McGraw-Hill, Inc., New York.

McCorkle, C. O. 1988 *Economics of Food Processing in the United States*, Academic Press, San Diego, CA.

McGee, H. 1984 *On Food and Cooking*, Scribner, New York.

Mellor, J. D. 1978 *Fundamentals of Freeze-Drying*, Academic Press, London.

Mercier, C., P. Linko, and J. M. Harper, Eds. 1989 *Extrusion Cooking*, American Association of cereal Chemists, St. Paul, MN.

Merrow, E. W., S. W. Chapel, and C. Worthing 1979 *A Review of Cost Estimation in New Technologies*, R-2481-DOE, The Rand Corporation, Santa Monica, CA.

Merrow, E. W., K. E. Phillips, and C. W. Myers 1981 *Understanding Cost Growth and Performance Shortfalls in Pioneer Process Plants*, R-2569-DOE, The Rand Corporation, Santa Monica, CA.

Merrow, E. W. 1989 *An Analysis of Cost Improvement in Chemical Process Technologies*, R-3357-DOE, The Rand Corporation, Santa Monica, CA.

Myers, C. W., R. F. Shangraw, M. R. Devey, and T. Hayashi 1986 *Understanding Process Plant Schedule Slippage and Startup Costs*, R-3215-PSSP/RC, The Rand Corporation, Santa Monica, CA.

Okos, M. R., Ed. 1986 *Physical and Chemical Properties of Food*, American Society of Agricultural Engineers, St. Joseph, MI.

Ramaswamy, H. S. and R. P. Singh 1997 *Sterilization Process Engineering*, Chapter 2 in Valentas et al. 1997.

Rockland, L. B. and L. R. Beuchat, Eds. 1987 *Water Activity: Theory and Applications to Food*, Marcel Dekker, New York.

Schultz, G. A. 2000 *Conveyor Safety*, American Society of Safety Engineers, Des Plaines, IL.

Schwartzberg, H. 1983 *A Compilation of Readings and Problems in Food Engineering*, University of Massachusetts, Amherst, MA.

Seiberling, D. A. 1997 CIP Sanitary Process Design, in *Handbook of Food Engineering Practice*, K. J. Valentas, E. Rotstein, and R. P. Singh, Eds., CRC Press, Boca Raton, FL.

Singh, R. P. and D. R. Heldman 2001 *Introduction to Food Engineering*, 3rd ed., Academic Press, London, UK.

Steffe, J. F. and R. P. Singh 1997 *Pipeline Design for Newtonian and Non-Newtonian Fluids*, in Valentas et al. 1997.

Tetra Pak 1998 *The Orange Book*, Tetra Pak Processing Systems AB, Lund, Sweden.

Troller, J. H. 1983 *Sanitation in Food Processing*, Academic Press, London.

Turton, R., R. C. Bailie, W. B. Whiting, and J. A. Shaewitz 1998 *Analysis, Synthesis and Design of Chemical Processes*, Prentice Hall PTR, Upper Saddle River, NJ.

Valentas, K. J., L. Levine, and J. P. Clark 1991 *Food Processing Operations and Scale-Up*, Marcel Dekker, Inc., New York.

Valentas, K. J., E. Rotstein, and R. P. Singh, Eds. 1997 *Handbook of Food Engineering Practice*, CRC Press, Boca Raton, FL.
Valle-Riestra, J. F. 1983 *Project Evaluation in the Chemical Process Industries*, McGraw-Hill Book Company, New York.
Van Arsdel, W. B., M. J. Copley, and A. I. Morgan 1973 *Food Dehydration*, AVI Publishing Company, Westport, CT.

Index